Connected  Mathematics™

Kaleidoscopes, Hubcaps, and Mirrors

Symmetry and Transformations

Teacher's Edition

Glenda Lappan
James T. Fey
William M. Fitzgerald
Susan N. Friel
Elizabeth Difanis Phillips

Developed at Michigan State University

DALE SEYMOUR PUBLICATIONS®
MENLO PARK, CALIFORNIA

Connected Mathematics™ was developed at Michigan State University with financial support from the Michigan State University Office of the Provost, Computing and Technology, and the College of Natural Science.

This material is based upon work supported by the National Science Foundation under Grant No. MDR 9150217.

This project was supported, in part,
by the
National Science Foundation ·
Opinions expressed are those of the authors
and not necessarily those of the Foundation

The Michigan State University authors and administration have agreed that all MSU royalties arising from this publication will be devoted to purposes supported by the Department of Mathematics and the MSU Mathematics Education Enrichment Fund.

This book is published by Dale Seymour Publications®, an imprint of Addison Wesley Longman, Inc.

Dale Seymour Publications
2725 Sand Hill Road
Menlo Park, CA 94025
Customer Service: 800-872-1100

Managing Editor: Catherine Anderson
Project Editor: Stacey Miceli
Book Editor: Mali Apple
ESL Consultant: Nancy Sokol Green
Production/Manufacturing Director: Janet Yearian
Production/Manufacturing Coordinator: Claire Flaherty
Design Manager: John F. Kelly
Photo Editor: Roberta Spieckerman
Design: Don Taka
Composition: London Road Design, Palo Alto, CA
Illustrations: Pauline Phung, Margaret Copeland, Ray Godfrey
Cover: Ray Godfrey

Photo Acknowledgements: 6 © Jeffry Myers/Stock, Boston; 7 (left) © Bob Kalman/The Image Works; 7 (center and right) © Mali Apple; 8 (left) © Photoworld/FPG International; 8 (center) © Science Source/Photo Researchers, Inc.; 8 (right) © Mali Apple; 9 © Charles Orrico/Superstock, Inc.; 11 © Bob Daemmrich/The Image Works; 14 (left) © Robert Frerck/Tony Stone Images; 14 (right) © Ira Kirschenbaum/Stock, Boston; 26 © Jean-Claude Lejeune/Stock, Boston; 38 © Richard B. Hoit/Photo Researchers, Inc.;

This Book is Printed
on Recycled Paper

Order number 21484
ISBN 1-57232-189-X

2 3 4 5 6 7 8 9 10-ML-01 00 99 98 97

The Connected Mathematics Project Staff

Project Directors

James T. Fey
University of Maryland

William M. Fitzgerald
Michigan State University

Susan N. Friel
University of North Carolina at Chapel Hill

Glenda Lappan
Michigan State University

Elizabeth Difanis Phillips
Michigan State University

Project Manager

Kathy Burgis
Michigan State University

Technical Coordinator

Judith Martus Miller
Michigan State University

Collaborating Teachers/Writers

Mary K. Bouck
Portland, Michigan

Jacqueline Stewart
Okemos, Michigan

Curriculum Development Consultants

David Ben-Chaim
Weizmann Institute

Alex Friedlander
Weizmann Institute

Eleanor Geiger
University of Maryland

Jane Miller
University of Maryland

Jane Mitchell
University of North Carolina at Chapel Hill

Anthony D. Rickard
Alma College

Evaluation Team

Mark Hoover
Michigan State University

Diane V. Lambdin
Indiana University

Sandra K. Wilcox
Michigan State University

Judith S. Zawojewski
National-Louis University

Graduate Assistants

Scott J. Baldridge
Michigan State University

Angie S. Eshelman
Michigan State University

M. Faaiz Gierdien
Michigan State University

Jane M. Keiser
Indiana University

Angela S. Krebs
Michigan State University

James M. Larson
Michigan State University

Ronald Preston
Indiana University

Tat Ming Sze
Michigan State University

Sarah Theule-Lubienski
Michigan State University

Jeffrey J. Wanko
Michigan State University

Field Test Production Team

Katherine Oesterle
Michigan State University

Stacey L. Otto
University of North Carolina at Chapel Hill

Teacher/Assessment Team

Kathy Booth
Waverly, Michigan

Anita Clark
Marshall, Michigan

Julie Faulkner
Traverse City, Michigan

Theodore Gardella
Bloomfield Hills, Michigan

Yvonne Grant
Portland, Michigan

Linda R. Lobue
Vista, California

Suzanne McGrath
Chula Vista, California

Nancy McIntyre
Troy, Michigan

Mary Beth Schmitt
Traverse City, Michigan

Linda Walker
Tallahassee, Florida

Software Developer

Richard Burgis
East Lansing, Michigan

Development Center Directors

Nicholas Branca
San Diego State University

Dianne Briars
Pittsburgh Public Schools

Frances R. Curcio
New York University

Perry Lanier
Michigan State University

J. Michael Shaughnessy
Portland State University

Charles Vonder Embse
Central Michigan University

Field Test Coordinators

Michelle Bohan
Queens, New York

Melanie Branca
San Diego, California

Alecia Devantier
Shepherd, Michigan

Jenny Jorgensen
Flint, Michigan

Sandra Kralovec
Portland, Oregon

Sonia Marsalis
Flint, Michigan

William Schaeffer
Pittsburgh, Pennsylvania

Karma Vince
Toledo, Ohio

Virginia Wolf
Pittsburgh, Pennsylvania

Shirel Yaloz
Queens, New York

Student Assistants

Laura Hammond
David Roche
Courtney Stoner
Jovan Trpovski
Julie Valicenti
Michigan State University

Patricia Wagner
Holmes Middle School

Greg Williams
Gundry Elementary School

Lansing

Susan Bissonette
Waverly Middle School

Kathy Booth
Waverly East Intermediate School

Carole Campbell
Waverly East Intermediate School

Gary Gillespie
Waverly East Intermediate School

Denise Kehren
Waverly Middle School

Virginia Larson
Waverly East Intermediate School

Kelly Martin
Waverly Middle School

Laurie Metevier
Waverly East Intermediate School

Craig Paksi
Waverly East Intermediate School

Tony Pecoraro
Waverly Middle School

Helene Rewa
Waverly East Intermediate School

Arnold Stiefel
Waverly Middle School

Portland

Bill Carlton
Portland Middle School

Kathy Dole
Portland Middle School

Debby Flate
Portland Middle School

Yvonne Grant
Portland Middle School

Terry Keusch
Portland Middle School

John Manzini
Portland Middle School

Mary Parker
Portland Middle School

Scott Sandborn
Portland Middle School

Shepherd

Steve Brant
Shepherd Middle School

Marty Brock
Shepherd Middle School

Cathy Church
Shepherd Middle School

Ginny Crandall
Shepherd Middle School

Craig Ericksen
Shepherd Middle School

Natalie Hackney
Shepherd Middle School

Bill Hamilton
Shepherd Middle School

Julie Salisbury
Shepherd Middle School

Sturgis

Sandra Allen
Eastwood Elementary School

Margaret Baker
Eastwood Elementary School

Steven Baker
Eastwood Elementary School

Keith Barnes
Sturgis Middle School

Wilodean Beckwith
Eastwood Elementary School

Darcy Bird
Eastwood Elementary School

Bill Dickey
Sturgis Middle School

Ellen Eisele
Sturgis Middle School

James Hoelscher
Sturgis Middle School

Richard Nolan
Sturgis Middle School

J. Hunter Raiford
Sturgis Middle School

Cindy Sprowl
Eastwood Elementary School

Leslie Stewart
Eastwood Elementary School

Connie Sutton
Eastwood Elementary School

Traverse City

Maureen Bauer
Interlochen Elementary School

Ivanka Berskshire
East Junior High School

Sarah Boehm
Courtade Elementary School

Marilyn Conklin
Interlochen Elementary School

Nancy Crandall
Blair Elementary School

Fran Cullen
Courtade Elementary School

Eric Dreier
Old Mission Elementary School

Lisa Dzierwa
Cherry Knoll Elementary School

Ray Fouch
West Junior High School

Ed Hargis
Willow Hill Elementary School

Richard Henry
West Junior High School

Dessie Hughes
Cherry Knoll Elementary School

Ruthanne Kladder
Oak Park Elementary School

Bonnie Knapp
West Junior High School

Sue Laisure
Sabin Elementary School

Stan Malaski
Oak Park Elementary School

Jody Meyers
Sabin Elementary School

Marsha Myles
East Junior High School

Mary Beth O'Neil
Traverse Heights Elementary School

Jan Palkowski
East Junior High School

Karen Richardson
Old Mission Elementary School

Kristin Sak
Bertha Vos Elementary School

Mary Beth Schmitt
East Junior High School

Mike Schrotenboer
Norris Elementary School

Gail Smith
Willow Hill Elementary School

Karrie Tufts
Eastern Elementary School

Mike Wilson
East Junior High School

Tom Wilson
West Junior High School

Minnesota

Minneapolis

Betsy Ford
Northeast Middle School

New York

East Elmhurst

Allison Clark
Louis Armstrong Middle School

Dorothy Hershey
Louis Armstrong Middle School

J. Lewis McNeece
Louis Armstrong Middle School

Rossana Perez
Louis Armstrong Middle School

Merna Porter
Louis Armstrong Middle School

Marie Turini
Louis Armstrong Middle School

North Carolina

Durham

Everly Broadway
Durham Public Schools

Thomas Carson
Duke School for Children

Mary Hebrank
Duke School for Children

Bill O'Connor
Duke School for Children

Ruth Pershing
Duke School for Children

Peter Reichert
Duke School for Children

Elizabeth City

Rita Banks
Elizabeth City Middle School

Beth Chaundry
Elizabeth City Middle School

Amy Cuthbertson
Elizabeth City Middle School

Deni Dennison
Elizabeth City Middle School

Jean Gray
Elizabeth City Middle School

John McMenamin
Elizabeth City Middle School

Nicollette Nixon
Elizabeth City Middle School

Malinda Norfleet
Elizabeth City Middle School

Joyce O'Neal
Elizabeth City Middle School

Clevie Sawyer
Elizabeth City Middle School

Juanita Shannon
Elizabeth City Middle School

Terry Thorne
Elizabeth City Middle School

Rebecca Wardour
Elizabeth City Middle School

Leora Winslow
Elizabeth City Middle School

Franklinton

Susan Haywood
Franklinton Elementary School

Clyde Melton
Franklinton Elementary School

Louisburg

Lisa Anderson
Terrell Lane Middle School

Jackie Frazier
Terrell Lane Middle School

Pam Harris
Terrell Lane Middle School

Ohio

Toledo

Bonnie Bias
Hawkins Elementary School

Marsha Jackish
Hawkins Elementary School

Lee Jagodzinski
DeVeaux Junior High School

Norma J. King
Old Orchard Elementary School

Margaret McCready
Old Orchard Elementary School

Carmella Morton
DeVeaux Junior High School

Karen C. Rohrs
Hawkins Elementary School

Marie Sahloff
DeVeaux Junior High School

L. Michael Vince
McTigue Junior High School

Brenda D. Watkins
Old Orchard Elementary School

Oregon

Canby

Sandra Kralovec
Ackerman Middle School

Portland

Roberta Cohen
Catlin Gabel School

David Ellenberg
Catlin Gabel School

Sara Normington
Catlin Gabel School

Karen Scholte-Arce
Catlin Gabel School

West Linn

Marge Burack
Wood Middle School

Tracy Wygant
Athey Creek Middle School

Pennsylvania

Pittsburgh

Sheryl Adams
Reizenstein Middle School

Sue Barie
Frick International Studies Academy

Suzie Berry
Frick International Studies Academy

Richard Delgrosso
Frick International Studies Academy

Janet Falkowski
Frick International Studies Academy

Joanne George
Reizenstein Middle School

Harriet Hopper
Reizenstein Middle School

Chuck Jessen
Reizenstein Middle School

Ken Labuskes
Reizenstein Middle School

Barbara Lewis
Reizenstein Middle School

Sharon Mihalich
Reizenstein Middle School

Marianne O'Connor
Frick International Studies Academy

Mark Sammartino
Reizenstein Middle School

Washington

Seattle

Chris Johnson
University Preparatory Academy

Rick Purn
University Preparatory Academy

Contents

Students often have an intuitive understanding of symmetry. They recognize that a design is symmetric if some part of it is repeated in a regular pattern. Though students begin recognizing symmetric figures at an early age, the analytic understanding needed to confirm symmetry and to construct figures with given symmetries requires greater mathematical sophistication. *Kaleidoscopes, Hubcaps, and Mirrors,* the last geometry and measurement unit in the Connected Mathematics™ curriculum, helps students to refine their knowledge of symmetry.

Symmetry is commonly described in terms of transformations. *Symmetry transformations,* or rigid motions, include reflections, rotations, and translations. These transformations are *congruences,* as opposed to the *similarities* discussed in the grade 7 unit *Stretching and Shrinking.* Similarity transformations change the size of a figure while preserving its shape. In contrast, symmetry transformations preserve both angle measures and side lengths, resulting in an image that is congruent to the original figure.

The purpose of this unit is to stimulate and sharpen students' awareness of symmetry and to begin to develop their understanding of the underlying mathematics. Students will explore congruence, symmetry, and transformations in greater depth in future mathematics classes.

In this unit, students study symmetry and symmetry transformations. They learn to recognize and create designs with symmetry, and they learn to describe mathematically the transformations that lead to symmetric designs.

Types of Symmetry

In the first investigation, students learn to recognize designs with symmetry and to identify lines of symmetry, centers and angles of rotation, and directions and lengths of translations.

A figure has *reflectional symmetry*, also called *mirror symmetry*, if a reflection over a line maps the figure exactly onto itself. The letter A below has reflectional symmetry because a reflection over the vertical line will match each point on the left half with a point on the right half. The vertical line is the *line of symmetry* for this design.

A figure has *rotational symmetry* if a rotation about a point maps the figure onto itself. The design below has rotational symmetry because a rotation of 120° or 240° about point *P* will match each flag with another flag. Point *P* is referred to as the *center of rotation*. The *angle of rotation* for this design is 120°, the smallest angle through which the design can be rotated to match with the original position.

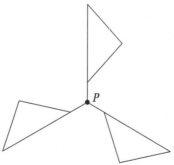

A figure has *translational symmetry* if a translation, or slide, maps the figure onto itself. The figure below is part of a translation-symmetric design. If this design continued in both directions, a slide of 1 inch to the right or left would match each flag in the design with another flag.

Creating Symmetric Designs

Once students learn to recognize symmetry in given designs, they create their own symmetric designs. Students may use reflecting devices, tracing paper, and angle rulers or protractors to help them construct such designs. A design with reflectional symmetry can be created by starting with a basic figure and then drawing the reflection of the figure over a line. A design with rotational symmetry can be created by starting with a basic figure and making n copies of the figure, where each copy is rotated $\frac{360}{n}$ degrees about a centerpoint from the previous copy. A figure with translational symmetry can be created by making copies of a basic figure so that each copy is a fixed distance and direction from the previous copy.

Symmetry Transformations

The concepts of symmetry are used as the starting point for the study of symmetry transformations, also called *distance-preserving transformations* or *rigid motions*. These transformations—reflections, rotations, and translations—match points with image points so that the distance between any two original points is equal to the distance between their images.

Students examine figures and their images under reflections, rotations, and translations, measuring key distances and angles. They use their findings to determine how they can specify a particular transformation so that another person could perform it exactly.

Students learn that a reflection can be specified by giving the line of reflection. They also write rules for describing reflections of figures drawn on a coordinate grid. Such rules tell how to find the image of a general point (x, y) under a reflection. For example, a reflection over the y-axis takes (x, y) to $(^-x, y)$; a reflection over the x-axis takes (x, y) to $(x, ^-y)$; and a reflection over the line $y = x$ takes (x, y) to (y, x).

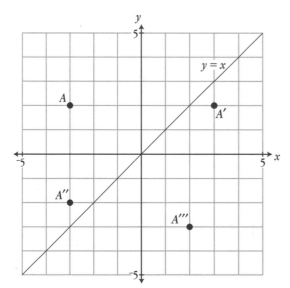

A' is the image of A under a reflection over the y-axis.

A'' is the image of A under a reflection over the x-axis.

A''' is the image of A under a reflection over the line $y = x$.

A rotation can be specified by giving the center of rotation and the angle of the turn. In this unit, the direction of the rotation is assumed to be counterclockwise unless a clockwise turn is specified. For example, a 45° rotation about a point P is a counterclockwise turn of 45° with P as the center of the rotation. As with reflections, students learn to specify certain rotations by giving rules for locating the image of a general point (x, y). For example, a rotation of 90° about the origin takes the point (x, y) to the image point $(\bar{y}, x)$, and a rotation of 180° about the origin takes (x, y) to $(\bar{x}, \bar{y})$.

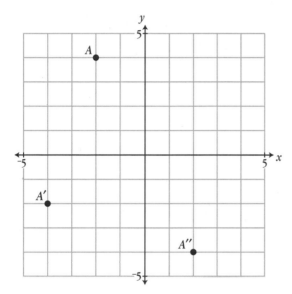

A′ is the image of A under a 90° rotation about the origin.

A″ is the image of A under a 180° rotation about the origin.

A translation can be specified by giving the length and direction of the slide. This can be done by drawing an arrow with the appropriate length and direction. A translation can also be specified by giving a rule for locating the image of a general point (x, y). For example, a translation of 3 units up takes point (x, y) to $(x, y + 3)$, and a translation of 3 units to the right takes (x, y) to $(x + 3, y)$. A translation along an oblique line can be specified by considering the vertical and horizontal components of the slide. For example, a translation of 2 units to the right and 4 units down takes (x, y) to $(x + 2, y - 4)$.

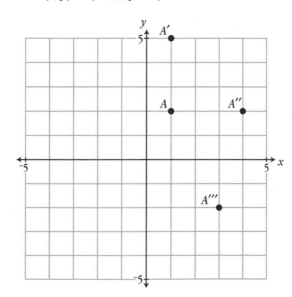

A′ is the image of A under a translation of 3 units up.

A″ is the image of A under a translation of 3 units to the right.

A‴ is the image of A under a under a translation of 2 units to the right and 4 units down.

A transformation matches every point in a plane with an image point. Although the points do not actually move, we often speak infomally of points being "moved" to new positions. (In fact, transformation geometry is often called *motion geometry*.) And it is not uncommon to focus on the effect of a transformation on a particular figure.

This unit attempts to give mathematically precise descriptions of transformations while accommodating students' natural instinct to visualize the figures moving. Thus, in many cases, students are asked to study a figure and its image without considering the effect of the transformation on other points. However, the "moved" figure is always referred to as the *image* of the original, and the vertices of the image are often labeled with primes or double primes to indicate that they are indeed different points.

Combining Transformations

Students explore combinations of transformations and try to find a single transformation that will give the same result as a given combination. For example, reflecting a figure over a line and then reflecting the image over a parallel line has the same result as translating the figure in a direction perpendicular to the reflection lines for a distance equal to twice that between the lines.

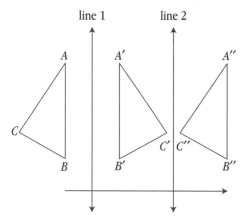

Reflecting a figure over a line and then reflecting the image over an intersecting line has the same result as rotating the original figure about the intersection point of the lines for an angle equal to twice that formed by the reflection lines.

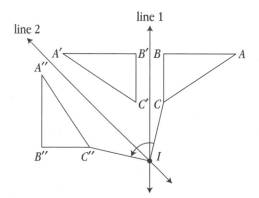

Students revisit this idea in the final investigation when they explore combinations of transformations that map a geometric figure onto itself.

Congruent Figures

The discussion of distance-preserving transformations leads naturally to the idea of congruence. Two figures are *congruent* if they have the same size and shape. Intuitively, this means that you could move one figure exactly on top of the other by a combination of rigid motions. In the language of transformations, two figures are congruent if there is a combination of distance-preserving transformations that maps one figure onto the other. Several problems ask students to explore this fundamental relationship among geometric figures.

Algebraic Properties of Transformation Combinations

In the last investigation of this unit, students explore tables showing combinations of symmetry transformations and relate the properties of the combining operation to the algebraic properties of real numbers. Students explore combinations of symmetry transformations on an equilateral triangle, a square, and a rectangle. For example, an equilateral triangle has three lines of symmetry and can be rotated by 120°, 240°, or 360° to coincide with its original position.

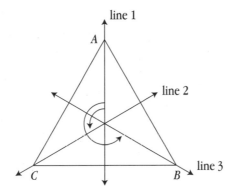

By preparing paper polygons and labeling the vertices on both sides, students create tools for studying combinations of symmetry transformations. The drawings below show how a paper triangle can be used to model a reflection over line 1 followed by a reflection over line 2. The result of the two reflections is the same as the result of a 120° rotation.

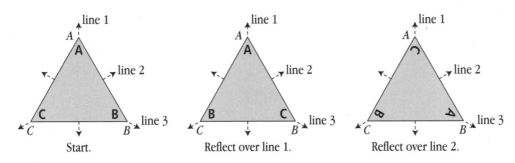

Students use their paper polygons to help them complete operation tables showing all the possible combinations of two symmetry transformations. The completed table for an equilateral triangle is shown below. The notation L_n means a reflection over line n, and the notation R_n means a rotation of n degrees. The symbol $*$, read as "and then," represents the combining operation. For example, the equation $L_1 * L_2 = R_{120}$ means that reflecting the triangle over line 1 "and then" reflecting it over line 2 is equivalent to rotating it 120°.

$*$	R_{360}	R_{120}	R_{240}	L_1	L_2	L_3
R_{360}	R_{360}	R_{120}	R_{240}	L_1	L_2	L_3
R_{120}	R_{120}	R_{240}	R_{360}	L_3	L_1	L_2
R_{240}	R_{240}	R_{360}	R_{120}	L_2	L_3	L_1
L_1	L_1	L_2	L_3	R_{360}	R_{120}	R_{240}
L_2	L_2	L_3	L_1	R_{240}	R_{360}	R_{120}
L_3	L_3	L_1	L_2	R_{120}	R_{240}	R_{360}

The set of symmetry transformations of an equilateral triangle under the combining operation forms a mathematical structure called a *group*. The properties exhibited in the table are as follows:

- The composition of any two symmetry transformations is equivalent to another symmetry transformation. This property is called *closure*.

- R_{360} is the *identity* element for the set of symmetry transformations under the combining operation. This means that for any symmetry transformation S, $S * R_{360} = R_{360} * S = S$. You can see this in the table by looking across the row labeled R_{360}. Each entry in this row is identical to the column head. Similarly, if you look down the column labeled R_{360}, you will see that each entry in the column is identical to the row head.

- Each transformation has an *inverse*. The inverse of a transformation is a transformation that, when combined with the original transformation in either order, gives the identity element. You can observe this in the table by noting that R_{360} occurs in every row and column. The symmetry transformations and their inverses for the equilateral triangle are

  R_{360} and R_{360}
  R_{120} and R_{240}
  L_1 and L_1
  L_2 and L_2
  L_3 and L_3

- Note that the combining operation is not commutative. For example, $R_{120} * L_1 = L_3$, but $L_1 * R_{120} = L_2$.

Technology

Connected Mathematics was developed with the belief that calculators should always be available and that students should decide when to use them.

If your students have access to computers, you may want to let them use a geometry or drawing program, such as Geometer's Sketchpad®, to explore transformations. The TesselMania!® and Tesselmania® Deluxe! programs are useful for exploring tessellations. These programs allow students to apply symmetry transformations to polygons to create tessellating shapes.

Mathematical and Problem-Solving Goals

Kaleidoscopes, Hubcaps, and Mirrors **was created to help students**

- Understand important properties of symmetry
- Recognize and describe symmetries of figures
- Use tools to examine symmetries and transformations
- Create figures with specified symmetries
- Identify basic design elements that can be used to replicate a given design
- Perform symmetry transformations of figures, including reflections, translations, and rotations
- Give precise mathematical directions for performing reflections, rotations, and translations
- Write coordinate rules for specifying the image of a general point (x, y) under particular transformations
- Combine transformations and find a single transformation that will produce the same result
- Find the symmetries of geometric figures and make tables showing the results of combining symmetry transformations
- Learn to appreciate the power of transformational geometry to describe motions, patterns, and designs in the real world

The overall goal of the Connected Mathematics curriculum is to help students develop sound mathematical habits. Through their work in this and other geometry units, students learn important questions to ask themselves about any situation that involves the principles explored in this unit, such as: *What makes an artistic design appealing? How can this design be described so that someone could re-create it? What patterns are apparent in the design? Can the patterns be used to make predictions about the design? What connections between geometry and algebra help create certain kinds of designs?*

Investigation 1: Three Types of Symmetry

Students are introduced to reflectional, rotational, and translational symmetry. They identify the symmetries in several designs and create designs with given symmetries. Students are also introduced to tools and procedures for testing for symmetry and making symmetric figures. The goal of this investigation is to heighten sensitivity to various forms of symmetry and to develop geometric techniques for testing and drawing symmetric figures.

Investigation 2: Symmetry Transformations

Students are challenged to describe the motions involved in constructing symmetric designs. They explore the relationships between figures and their images under reflections, rotations, and translations. They use their findings to write precise rules for finding images under each type of transformation. Students also look at combinations of reflections. They reflect a figure over intersecting lines and parallel lines. In each case, they determine whether the combination of reflections is equivalent to a single transformation.

Investigation 3: Transforming Coordinates

Students work with figures drawn on a coordinate grid. By writing computer commands for creating figures and their images under various transformations, students develop rules for locating the image of a general point (x, y) under a particular reflection, rotation, or translation. The last problem uses the ideas of rigid motions to introduce the notion of congruent figures.

Investigation 4: Symmetry and Algebra

Students explore combinations of symmetry transformations of an equilateral triangle and a square, and they create tables showing the results of every combination of two symmetry transformations for these figures. Students then use their tables to determine whether the combining operation satisfies important algebraic properties.

The ideas in *Kaleidoscopes, Hubcaps, and Mirrors* build on and connect to several big ideas in other Connected Mathematics units.

Big Idea	Prior Work	Future Work
recognizing symmetry in designs determining the design element that has been reflected, rotated, or translated to produce a design with symmetry creating designs with reflectional, rotational, or translational symmetries	recognizing and completing mirror reflections *(Shapes and Designs; Ruins of Montarek)* recognizing and completing designs with rotational symmetry *(Shapes and Designs)* rotating cube buildings *(Ruins of Montarek)* recognizing, analyzing, and producing tessellations *(Shapes and Designs)*	recognizing symmetry in graphs of functions *(high school)* applying the ideas of symmetry to other subjects, such as graphic design and architecture *(high school and college)*
looking for patterns that can be used to predict attributes of designs	looking for regularity and using patterns to make predictions *(all Connected Mathematics units)*	making inferences and predictions based on observation and proving predictions *(high school and college)*
relating rigid motions to the concept of symmetry	relating similarity transformations to the concept of similarity *(Stretching and Shrinking)*	describing symmetry in graphs, such as graphs of quadratic functions, periodic functions, and power functions *(high school)*
describing rigid motions in words and with coordinate rules	describing similarity transformations in words and with coordinate rules *(Stretching and Shrinking)*	using matrices to represent transformations *(high school)*
composing symmetry transformations	performing similarity transformations *(Stretching and Shrinking)*	composing functions; recognizing graphs of certain functions as transformations of graphs of other functions, such as recognizing that the graph of $f(x) = (x + 2)^2$ is the graph of $f(x) = x^2$ shifted 2 units to the left *(high school)*
making tables of combinations of symmetry transformations for geometric figures and exploring the group structure in the tables	making tables for real-number operations *(elementary school)* looking at the properties of real-number operations *(Say It with Symbols)*	studying the algebraic structure of groups, rings, and fields *(high school and college)*

Materials

For students

- Labsheets
- Paper copies of triangle *ABC* and square *ABCD* (cut from the blackline masters)
- Graphing calculators (preferably with the capacity to display a function as a table)
- Transparent reflection tools (such as Image Reflectors™)
- Mirrors
- Angle rulers or protractors (1 per student)
- Compasses (1 per student)
- Rulers (1 per student)
- Tracing paper (or lightweight plain paper)
- Centimeter and quarter-inch grid paper (provided as blackline masters)
- Transparent grids (optional; copy the grid onto transparency film)
- Blank transparencies (optional)
- Markers
- 1-inch-wide paper strips
- Paper copies of hexagon *ABCDEF* (cut from the blackline master) and isometric dot paper (optional; for the optional Unit Project)

For the teacher

- Transparencies and transparency markers (optional)
- Transparencies of Labsheets 1.ACE1, 1.ACE2, 1.ACE3, 1.ACE4, 2.1B, 2.2, 2.3B, 2.4B, 2.ACE1, 2.ACE2, 2.ACE3, 2.ACE4, 4.1A, 4.1B, 4.2A, 4.2B, and 4.ACE (optional)
- Transparent centimeter grids (optional; copy the grid onto transparency film)
- Blank transparencies (optional)
- Overhead display model of students' graphing calculators (optional)
- Computer graphics software (optional; for demonstration)
- Two mirrors (optional)
- Rubber stamp (optional)
- Transparent ruler (optional)
- Chalkboard compass (optional)

Resources

Books

Britton, Jill and Walter Britton, *Teaching Tessellating Art.* Palo Alto, Calif.: Dale Seymour Publications, 1992.

Finkel, Leslie, *Kaleidoscope Designs and How to Create Them.* New York: Dover Publications, 1980.

Seymour, Dale and Jill Britton, *Introduction to Tessellations.* Palo Alto, Calif.: Dale Seymour Publications, 1992.

Software Packages

Geometer's Sketchpad (Macintosh and Windows). Emeryville, Calif.: Key Curriculum Press.

TesselMania Deluxe! (Macintosh and Windows). Minneapolis: Minnesota Educational Computing Corporation.

Posters

Desoe, Carol. *Symmetry in Wheels.* Palo Alto, Calif.: Dale Seymour Publications, 1994.

Kim, Scott. *Inversions.* Palo Alto, Calif.: Dale Seymour Publications, 1981.

Seymour, Dale. *Kaleidoscope Symmetry.* Palo Alto, Calif.: Dale Seymour Publications, 1993.

Vojack, Robert. *Symmetry in Fractals.* Palo Alto, Calif.: Dale Seymour Publications, 1994.

Pacing Chart

This pacing chart gives estimates of the class time required for each investigation and assessment piece. Shaded rows indicate opportunities for assessment.

Investigations and Assessments	Class Time
1 Three Types of Symmetry	4 days
Check-Up 1	$\frac{1}{2}$ day
2 Symmetry Transformations	6 days
Quiz	1 day
3 Transforming Coordinates	6 days
Check-Up 2	$\frac{1}{2}$ day
4 Symmetry and Algebra	4 days
Self-Assessment	Take home
Unit Test	1 day
Unit Project	Take home

Kaleidoscopes, Hubcaps, and Mirrors Vocabulary

The following words and concepts are used in *Kaleidoscopes, Hubcaps, and Mirrors.* Concepts in the left column are those essential for student understanding of this and future units. The Descriptive Glossary gives descriptions of many of these terms.

Essential terms developed in this unit	Terms developed in previous units	Nonessential terms
congruent figures	diagonal	angle of rotation
line reflection	equilateral triangle	center of rotation
reflectional symmetry	hexagon	commutative operation
rotation	parallel	direction of translation
rotational symmetry	parallelogram	identity element
symmetry	perpendicular	image
transformation	tessellation	inverse element
translation		kaleidoscope
translational symmetry		line of symmetry
		reflection line
		strip pattern

Assessment Summary

Embedded Assessment

Opportunities for informal assessment of student progress are embedded throughout *Kaleidoscopes, Hubcaps, and Mirrors* in the problems, the ACE questions, and the Mathematical Reflections. Suggestions for observing as students explore and discover mathematical ideas, for probing to guide their progress in developing concepts and skills, and for questioning to determine their level of understanding can be found in the Launch, Explore, and Summarize sections of all investigation problems. Some examples:

- Investigation 2, Problem 2.3 *Launch* (page 41f) suggests a demonstration you can use to introduce your students to the idea of specifying the image of an object under a rotation.

- Investigation 1, Problem 1.3 *Explore* (page 23g) suggests an extension activity for students who are interested in exploring the mathematics of kaleidoscopes.

- Investigation 3, Problem 3.2 *Summarize* (page 58e) suggests a discussion you can use to assess and extend your students' understanding of writing coordinate rules for specifying the results of translations.

ACE Assignments

An ACE (Applications—Connections—Extensions) section appears at the end of each investigation. To help you assign ACE questions, a list of assignment choices is given in the margin next to the reduced student page for each problem. Each list indicates the ACE questions that students should be able to answer after they complete the problem.

Check-Ups

Two check-ups, which may be given after Investigations 1 and 3, are provided for use as quick quizzes or warm-up activities. The check-ups are designed for students to complete individually. You will find the check-ups and their answer keys in the Assessment Resources section.

Partner Quiz

One quiz, which may be given after Investigation 2, is provided with this unit. The quiz is designed to be completed by pairs of students with the opportunity for revision based on teacher feedback. You will find the quiz and its answer key in the Assessment Resources section. As an alternative to the quiz provided, you can construct your own quiz by combining questions from the Question Bank, this quiz, and unassigned ACE questions.

Question Bank

A Question Bank provides questions you can use for homework, reviews, or quizzes. You will find the Question Bank and its answer key in the Assessment Resources section.

Notebook/Journal

Students should have notebooks to record and organize their work. Notebooks should include student journals and sections for vocabulary, homework, quizzes, and check-ups. In their journals, students can take notes, solve investigation problems, and record their ideas about Mathematical Reflections questions. Journals should be assessed for completeness rather than correctness; they should be seen as "safe" places where students can try out their thinking. A Notebook Checklist and a Self-Assessment are provided in the Assessment Resources section. The Notebook Checklist helps students organize their notebooks. The Self-Assessment guides students as they review their notebooks to determine which ideas they have mastered and which they still need to work on.

The Unit Test

As the final assessment for this unit, you may assign the unit project or administer the unit test. The test focuses on describing symmetries and transformations, performing transformations and combinations of transformations, and making tables of transformation combinations.

The Unit Project

As the final assessment for this unit, you may assign the unit project or administer the unit test. In part 1 of the project, students create tessellating shapes from a square and a rhombus. To make a tessellating shape, students cut pieces from the sides of a square or rhombus and rotate them about a vertex. In part 2, students make an origami wreath and a pinwheel and analyze the symmetries in these shapes. A guide to the project is included in the Assessment Resources section.

Introducing Your Students to *Kaleidoscopes, Hubcaps, and Mirrors*

Students' natural fascination with symmetric patterns is used to introduce this unit.

To find out how much students already know about symmetry, direct their attention to the objects pictured on the opening pages of the student edition. Ask them to look at first the butterfly, then the hubcap, and then the quilt pattern and to observe how part of the object can be folded, rotated, or slid to match another part. For each object, ask students whether they see other objects—on the page or somewhere in the classroom—that show the same property. Ask them to describe, as clearly as they can, each of the three properties.

Explain to the class that all of these objects have symmetry. You might end by asking, "What do you think the word *symmetry* means?" and allowing students to share their ideas.

Kaleidoscopes, Hubcaps, and Mirrors

Look at the drawing of the butterfly. The right half of the butterfly is the mirror image of the left half. You could fold the butterfly so that the two halves match exactly. Does anything else on these pages show this property? Can you think of other things that have this property?

Look at the hubcap. You could rotate the hubcap less than a full turn to a point at which it looks the same as the original. Does anything else on these pages show this property? Can you think of other things that have this property?

Now, look at the pattern on the quilt. You could slide one part of the pattern in a straight line so that it exactly matches another part. Does anything else on these pages show this property? Can you think of other things that have this property?

The butterfly, the hubcap, and the quilt pattern that you see here may look very different, but they share an important characteristic called *symmetry*. There is some part of each drawing that is repeated to create a balanced pattern. In each drawing, can you describe the part and how it is repeated to make the pattern?

In this unit, you will study the mathematical properties of different types of symmetry. You will also investigate the motions that are used to create symmetric designs.

Mathematical Highlights

The Mathematical Highlights page provides information for students and for parents and other family members. It gives students a preview of the activities and problems in *Kaleidoscopes, Hubcaps, and Mirrors.* As they work through the unit, students can refer back to the Mathematical Highlights page to review what they have learned and to preview what is still to come. This page also tells students' families what mathematical ideas and activities will be covered as the class works through *Kaleidoscopes, Hubcaps, and Mirrors.*

Mathematical Highlights

In *Kaleidoscopes, Hubcaps, and Mirrors,* you will learn about symmetry and symmetry transformations.

● As you look at hubcaps, kaleidoscope designs, and tessellations, you discover important properties of reflectional, rotational, and translational symmetries.

● By using mirrors, tracing paper, and other tools, you create designs with various types of symmetry.

● Examining figures and their images under different types of transformations helps you develop mathematical definitions for rotation, reflection, and translation.

● Applying one transformation to a figure and a second transformation to its image leads you to discover that combining two transformations is often equivalent to applying a single transformation.

● Writing computer commands for drawing the images of figures under various transformations leads you to develop rules for transforming figures on a coordinate grid.

● As you explore combinations of symmetry transformations for regular polygons, you discover that the combining operation satisfies some important algebraic properties.

● By applying all that you have learned about symmetry and transformations, you create tessellations, wreaths, and pinwheels.

Using a Calculator

As you work on the Connected Mathematics™ units, you decide whether to use a calculator to help you solve a problem.

The Investigations

The teaching materials for each investigation consist of three parts: an overview, student pages with teaching outlines, and detailed notes for teaching the investigation.

The overview of each investigation includes brief descriptions of the problems, the mathematical and problem-solving goals of the investigation, and a list of necessary materials.

Essential information for teaching the investigation is provided in the margins around the student pages. The "At a Glance" overviews are brief outlines of the Launch, Explore, and Summarize phases of each problem for reference as you work with the class. To help you assign homework, a list of "Assignment Choices" is provided next to each problem. Where space permits, answers to problems, follow-ups, ACE questions, and Mathematical Reflections appear next to the appropriate student pages.

The Teaching the Investigation section follows the student pages and is the heart of the Connected Mathematics curriculum. This section describes in detail the Launch, Explore, and Summarize phases for each problem. It includes all the information needed for teaching, along with suggestions for what you might say at key points in the teaching. Use this section to prepare lessons and as a guide for teaching the investigations.

Assessment Resources

The Assessment Resources section contains blackline masters and answer keys for the check-ups, the quiz, the Question Bank, and the Unit Test. Blackline masters for the Notebook Checklist and the Self-Assessment are given. These instruments support student self-evaluation, an important aspect of assessment in the Connected Mathematics curriculum.

Blackline Masters

The Blackline Masters section includes masters for all labsheets and transparencies. Blackline masters of geometric figures, centimeter and quarter-inch grid paper, and isometric dot paper are also provided.

Additional Practice

Practice pages for each investigation offer additional problems for students who need more practice with the basic concepts developed in the investigations as well as some continual review of earlier concepts.

Descriptive Glossary

The glossary provides descriptions and examples of the key concepts in *Kaleidoscopes, Hubcaps, and Mirrors*. These descriptions are not intended to be formal definitions but are meant to give you an idea of how students might make sense of these important concepts.

Three Types of Symmetry

This first investigation will familiarize students with the three basic types of symmetry: reflectional, rotational, and translational. Students will investigate ways to identify the symmetry in a design and methods for creating designs that have symmetry. They will make use of various tools for finding symmetry and creating symmetric designs. Students will also be introduced to the idea of determining a *basic design element*—a piece of a pattern or design that, when transformed using at least one type of symmetry transformation, will create the entire design.

In Problem 1.1, Reflectional Symmetry, students explore ways for checking for reflectional symmetry and search for examples of such symmetry in several designs. In Problem 1.2, Rotational Symmetry, they analyze illustrations of hubcaps to explore the properties of rotational symmetry and the concept of angle of rotation. In Problem 1.3, Symmetry in Kaleidoscope Designs, students look for reflectional and rotational symmetry in various kaleidoscope designs and are introduced to the idea of a basic design element. In Problem 1.4, Translational Symmetry, they begin to look at tessellations to find translational and other types of symmetry and to further explore the idea of a basic design element.

Mathematical and Problem-Solving Goals

- **To explore reflectional, rotational, and translational symmetry informally**

- **To explore the use of tools, such as tracing paper, to analyze designs to determine their symmetries**

- **To design shapes that have specified symmetries**

- **To identify basic design elements that can be used to replicate a design**

Materials

Problem	For students	For the teacher
All	Graphing calculators, tracing paper (or lightweight plain paper), rulers, mirrors, transparent reflection tools such as Image Reflectors	Transparencies: 1.1 to 1.4 (optional), overhead graphing calculator (optional)
1.1	Labsheet 1.1 (1 per student)	
1.2	Labsheet 1.2 (1 per student), angle rulers or protractors	
1.3	Labsheet 1.3 (1 per student), markers, angle rulers or protractors	Two mirrors (optional)
1.4	Labsheet 1.4 (1 per student), markers, angle rulers or protractors, 1-inch-wide paper strips, blank transparencies (optional)	Rubber stamp (optional)
ACE	Labsheets 1.ACE1, 1.ACE2, 1.ACE3, and 1.ACE4 (1 each per student); tracing paper; compasses (1 per student)	Transparencies of Labsheets 1.ACE1, 1.ACE2, 1.ACE3, and 1.ACE4 (optional)

Three Types of Symmetry

When part of an object or design is repeated to create a balanced pattern, we say that the object or design has *symmetry*. Butterflies, the blades of a fan, and wallpaper designs often have symmetry.

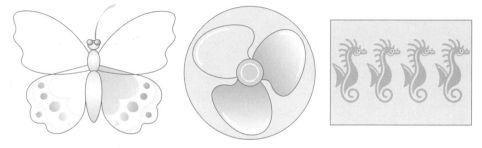

You can find examples of symmetry all around you. Artists use symmetry to make designs that are pleasing to the eye. Architects use symmetry to create a sense of balance in their buildings. And symmetry is found throughout the natural world.

There are several types of symmetry:

- The butterfly above has **reflectional symmetry,** which is also called *line symmetry* or *mirror symmetry*. If you drew a vertical line through the center of the butterfly, the two halves of the drawing would be mirror images.

- The fan blades have **rotational symmetry**. You could rotate the drawing less than a full turn about its centerpoint to a position in which it looks the same as the original drawing.

- The wallpaper design has **translational symmetry**. You could slide, or *translate,* any figure in the design in a regular straight-line pattern so that it exactly matches any other figure.

These three types of symmetry have unique mathematical properties. In this investigation, you will look for symmetry, and you will create designs with different types of symmetry.

Reflectional Symmetry

Launch

- Introduce reflectional, rotational, and translational symmetry.

- Talk about how students might look for reflectional symmetry, and demonstrate the use of any available tools.

- Have individuals begin the problem and then gather in pairs to finish it and the follow-up.

Explore

- Encourage accuracy as students draw lines of symmetry.

- Make sure students are looking for symmetry of the *entire* design.

Summarize

- Have students share the lines of symmetry they found and the methods they used.

You may already have experience creating shapes and designs with *reflectional symmetry*. You can produce a symmetric design by folding a sheet of paper in half and making cuts.

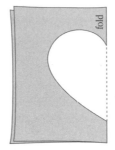

The fold through the center of the heart above is the line of symmetry. A *line of symmetry* divides a design into halves that are mirror images. If you fold a design on a line of symmetry, the halves will exactly match. Mirrors and other reflecting devices are useful for checking for reflectional symmetry. When you place a mirror on a line of symmetry, the reflected image looks identical to the design behind the mirror.

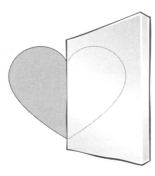

Tracing paper can also be helpful in looking for reflectional symmetry. You can trace one half of a design and then flip your tracing over the line of symmetry to see whether it matches the other half of the design.

Assignment Choices

ACE questions 4–7 and 18a–18e

Answers to Problem 1.1

A.

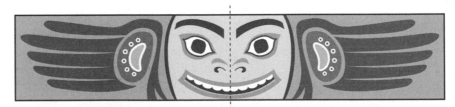

B.

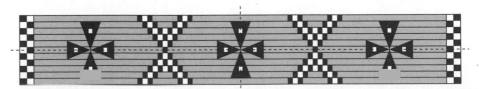

Think about this!

The objects below have reflectional symmetry. Can you identify the lines of symmetry?

Problem 1.1

The designs below are reproduced on Labsheet 1.1. Use mirrors, tracing paper, or other tools to help you find all the lines of symmetry in each design.

A.

B.

C.

D.

C.

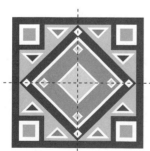

Note: Students may think there are two additional lines of symmetry on the diagonals; the orientation of the eight small triangles makes these incorrect.

D.   There are no lines of symmetry in this design.

Rotational Symmetry

Launch

- Introduce the ideas of rotational symmetry and angle of rotation.

- Have individuals do the problem and then gather in pairs to share ideas.

Explore

- Ask students who find the angles of rotation easily if they can find them without measuring.

- Have students who finish early do the follow-up.

Summarize

- Have students share their findings and methods.

- Explore the relationship between the angle of rotation and the number of rotation angles.

Assignment Choices

ACE questions 1–3, 13–17, 18f, 18g, and unassigned choices from earlier problems

---

■ **Problem 1.1 Follow-Up**

Explain why these traditional quilt designs do *not* have reflectional symmetry.

1.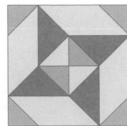

2.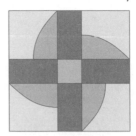

Next-door neighbor   Nocturne

1.2 **Rotational Symmetry**

The pinwheel design at right has *rotational symmetry*. It can be rotated 45°, 90°, 135°, 180°, 225°, 270°, 315°, or 360° about its centerpoint to a position in which it looks the same as the original design. The *angle of rotation* is the smallest angle through which a design can be rotated to coincide with the original design. The angle of rotation for this pinwheel is 45°. Notice that the other rotation angles are multiples of 45°.

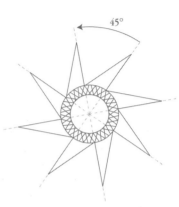

Think about this!

The objects below display rotational symmetry. Can you identify the angles of rotation?

Answers to Problem 1.1 Follow-Up

1, 2. Explanations will vary. Because the patterns are squares, the only possible lines of symmetry are a horizontal line through the center, a vertical line through the center, and along the two diagonals. However, none of these are lines of symmetry for these designs, a fact that can be confirmed by any of several methods.

Answers to Problem 1.2

A. The angles of rotation for hubcaps 1, 2, 3, and 4 are 72°, 120°, 36°, and 40°, respectively. Explanations will vary.

Problem 1.2

Rotational symmetry can be found in many objects that rotate about a centerpoint. For example, the automobile hubcaps shown below have rotational symmetry. These hubcaps are reproduced on Labsheet 1.2.

A. Determine the angle of rotation for each hubcap. Explain how you found the angle.

B. Some of the hubcaps also have reflectional symmetry. Sketch all the lines of symmetry for each hubcap.

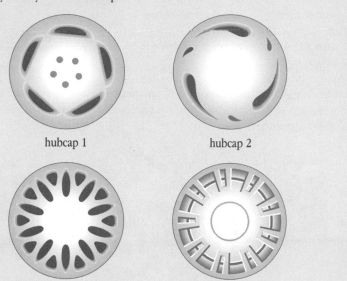

hubcap 1   hubcap 2

hubcap 3   hubcap 4

■ Problem 1.2 Follow-Up

1. Create a hubcap design that has rotational symmetry with a 90° angle of rotation but no reflectional symmetry.

2. Create a hubcap design that has rotational symmetry with a 60° angle of rotation and at least one line of symmetry.

3. Why do you think many rotating objects are designed to have rotational symmetry?

Investigation 1: Three Types of Symmetry `9`

B.

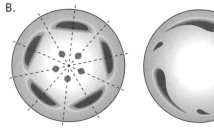 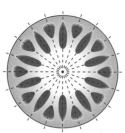 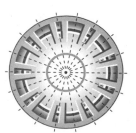 

hubcap 1   hubcap 2   hubcap 3   hubcap 4

Answers to Problem 1.2 Follow-Up

See page 23k.

Symmetry in Kaleidoscope Designs

At a Glance

Grouping:
pairs

Launch

■ As a class, look at the kaleidoscope designs that are pictured in the student edition.

■ Have pairs explore the problem and follow-up.

Explore

■ Encourage students to use tools to find all the reflectional and rotational symmetries.

Summarize

■ Have students share and discuss the symmetries they found.

■ Help the class see the mathematical connection between reflectional and rotational symmetry.

Assignment Choices

ACE questions 25–33 and unassigned choices from earlier problems

---

1.3 Symmetry in Kaleidoscope Designs

A **kaleidoscope** is a tube containing colored beads or pieces of glass and carefully placed mirrors. When a kaleidoscope is held to the eye and rotated, the viewer sees colorful, symmetric patterns.

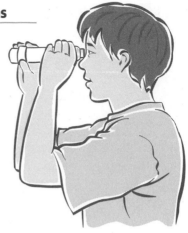

Kaleidoscopes were patented in 1817 by the Scottish scientist Sir David Brewster. Brewster was intrigued by the science of nature, and he developed kaleidoscopes in an effort to simulate the designs he saw in the world around him.

The designs below are called *kaleidoscope designs* because they are similar to designs you would see if you looked through a kaleidoscope. In this problem, you will explore the symmetries in these designs.

design 1   design 2   design 3

design 4   design 5   design 6

Source: Dale Seymour. *Kaleidoscope Symmetry* poster. Palo Alto, Calif.: Dale Seymour Publications, 1993.

10   Kaleidoscopes, Hubcaps, and Mirrors

Answers to Problem 1.3

A.   See page 23l.

B.   Designs 1, 2, 3, and 6 have a 120° angle of rotation. Designs 4 and 5 have a 60° angle of rotation.

Problem 1.3

The kaleidoscope designs are reproduced on Labsheet 1.3. Do parts A and B for each design.

A. Look for reflectional symmetry in each design. Sketch all the lines of symmetry you find.

B. Look for rotational symmetry in each design. Determine the angle of rotation for each design.

■ Problem 1.3 Follow-Up

1. For the kaleidoscope designs in Problem 1.3, describe the relationship between the number of lines of symmetry and the angle of rotation.

2. For each kaleidoscope design, sketch or outline the *basic design element*. That is, sketch or outline a part of the design from which the entire design could be created.

Answers to Problem 1.3 Follow-Up

1. Excluding design 3, which has no reflectional symmetry, the product of the number of lines of symmetry and the angle of rotation is always 360°: 3 × 120° = 360° and 6 × 60° = 360°.

2. See page 23l.

Translational Symmetry

At a Glance

Grouping:
small groups

Launch

- Talk about the idea of using a basic design element and an arrow to specify how to re-create part of a pattern.

- Have groups of three or four explore the problem and follow-up.

Explore

- Distribute blank transparencies for students to copy their design elements. *(optional)*

- Ask students whether there are other ways to specify each design element or translation.

Summarize

- Have students share basic design elements and translations for each tessellation.

Assignment Choices

ACE questions 8–12, 19–24, and unassigned choices from earlier problems

Assessment

It is appropriate to use Check-Up 1 after this problem.

1.4 Translational Symmetry

A **translation** is a geometric motion that slides a figure from one position to another. A design has *translational symmetry* if it can be created by sliding a basic design element in a regular, straight-line pattern. Designs on ribbons, belts, and wallpaper often have translational symmetry.

To describe the translational symmetry in a design, you can draw the basic design element and an arrow indicating the direction and length of a slide that would move one part of the design to another.

Problem 1.4

The designs below are tessellations. A **tessellation** is a design made from copies of a single basic design element that cover a surface without gaps or overlaps. The tessellations are reproduced on Labsheet 1.4. Each tessellation has translational symmetry. Do parts A and B for each tessellation.

A. Outline a basic design element that could be used to create the tessellation using only translations.

B. Write directions or draw an arrow showing how the basic design element can be copied and slid to produce another part of the pattern.

tessellation 1

tessellation 2

tessellation 3

tessellation 4

Answers to Problem 1.4

A, B. There are numerous ways to identify a basic design element and to specify a translation for each basic design element; a few are shown here for each tessellation.

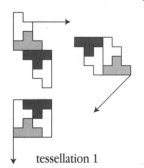

tessellation 1

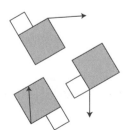

tessellation 2

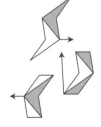

tessellation 3

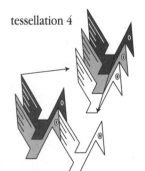

tessellation 4

■ **Problem 1.4 Follow-Up**

1. Describe the rotational and reflectional symmetries that appear in the tessellations in Problem 1.4. As you consider possible symmetries, imagine that the tessellations continue forever in all directions.

2. Cut a strip of paper about 1 inch wide. Make a design with translational symmetry on your strip by continuing the following pattern. This pattern appears in fabrics created by the Mayan people of Central America.

Think about this!

Can you identify the symmetries in the designs below?

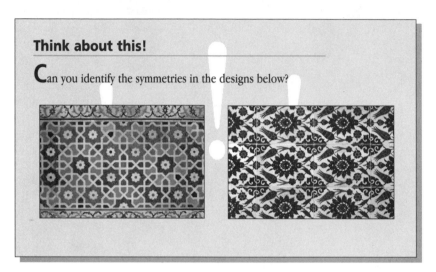

Answers to Problem 1.4 Follow-Up

1. Answers will vary, depending on whether students take the coloring in each design into account. Disregarding the coloring, tessellations 1 and 2 have rotational symmetry with a 90° angle of rotation, and tessellation 3 has rotational symmetry with a 180° angle of rotation. Tessellation 4 has no other type of symmetry.

2.

Applications • Connections • Extensions

As you work on these ACE questions, use your calculator whenever you need it.

Applications

In 1–3, a line design is given. A *line design* is a pattern that is formed entirely by straight lines but that appears to contain curves. On Labsheet 1.ACE1, indicate the lines of symmetry and the angle of rotation for the design.

1.

2.

3.

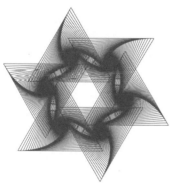

Source: Dale Seymour. *Introduction to Line Designs* Palo Alto, Calif.: Dale Seymour Publications, 1992.

Answers

Applications

1. This design has rotational symmetry about the centerpoint with a 36° angle of rotation.

2. This design has reflectional symmetry about a horizontal line through the center, reflectional symmetry about a vertical line through the center, and rotational symmetry about the centerpoint with a 180° angle of rotation.

3. This design has rotational symmetry about the centerpoint with a 60° angle of rotation.

4.

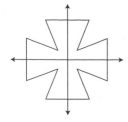

5. See below right.

6.

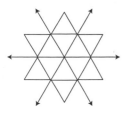

7. See below right.

8. See page 23m.

In 4–7, a basic design element and one or more lines are given. On Labsheet 1.ACE2, use the basic design element to create a design with the given lines of symmetry.

4.

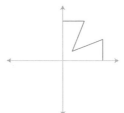

5. COOKIE

6.

7. 1 2 3 4 5 6 7 8 9 0

8. A tessellation of stars and hexagons has been started on the triangular grid below. Labsheet 1.ACE2 contains a copy of this figure. Continue the pattern to completely fill the grid.

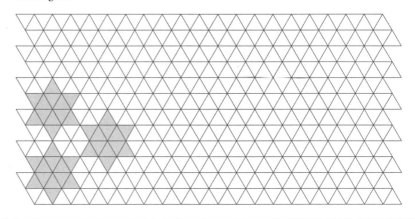

5. COOKIE

7. 1 2 3 4 5 6 7 8 9 0

a. Describe all the types of symmetry in this tessellation, and give at least one example of each type. As you consider possible symmetries, imagine that the tessellation continues forever in all directions.

b. Describe how you could check each symmetry by using a mirror or tracing paper or by folding the labsheet.

In 9–12, do parts a–c. These tessellations are reproduced on Labsheet 1.ACE3.

9.

10.

11.

12.

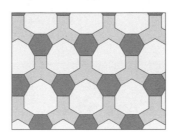

a. Sketch a basic design element that could be used to create the tessellation using only translations.

b. Tell what types of symmetry appear in the tessellation. Imagine that each tessellation continues forever in all directions.

c. Describe the lines of symmetry for the reflectional symmetries, the centers and angles of rotation for the rotational symmetries, and the directions and lengths of the translational symmetries.

9c. There are vertical lines of symmetry through the centers of the dark pieces and horizontal lines of symmetry through the centers of the light pieces. There is 180° rotational symmetry about the point where any two dark pieces meet. There are many translational symmetries including a vertical slide, a horizontal slide, and a diagonal slide. Some lines of symmetry, centers of rotation, and translations are shown below.

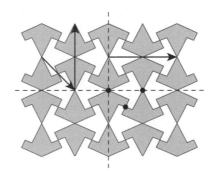

9b. This tessellation has reflectional, rotational, and translational symmetries.

9c. See below left.

10a. Possible answer:

10b. This pattern has reflectional, rotational, and translational symmetries.

10c. See page 23m.

11a. Possible answer:

11b. This pattern has rotational and translational symmetries.

11c. See page 23n.

12a. Possible answer:

12b. This pattern has reflectional, rotational, and translational symmetries.

12c. See page 23n.

13. *Zane* has reflectional symmetry with a vertical line of symmetry between the *a* and the *n*.

14. *Yvette* has reflectional symmetry with a vertical line of symmetry through the center of the letters.

15. *Michael* has rotational symmetry with a center of rotation in the middle of the design and a 180° angle of rotation.

16. *Eve* has reflectional symmetry with a vertical line of symmetry through the center of the *V*.

17. *Quincy* has rotational symmetry with a center of rotation in the middle of the design and a 180° angle of rotation.

Connections

18a. See below right.

In 13–17, describe the symmetry in the name.

13.

14.

15.

16.

17.

Names © 1984 Scott Kim.

Connections

18. Many of the letters in our alphabet have symmetry.

A B C D E F G H I J K L M
N O P Q R S T U V W X Y Z

   a. Identify the capital letters that have reflectional symmetry. Sketch each letter you find, and show all the lines of symmetry.

18a.   A B C D E H I K
          M O T U V W X Y

b. What state name has reflectional symmetry when written horizontally in capital letters?

c. What four state names have reflectional symmetry when written vertically in capital letters?

d. Write at least one word, name, or phrase that has reflectional symmetry when written horizontally.

e. Write at least one word, name, or phrase that has reflectional symmetry when written vertically.

f. Identify the capital letters that have rotational symmetry. Sketch each letter you find, and give its angle of rotation.

g. Write at least one word, name, or phrase that has rotational symmetry.

19. In this investigation, you studied designs with reflectional, rotational, and translational symmetry. The design below is a bit different from those you have seen.

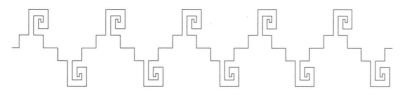

a. Trace a basic design element from which the rest of the pattern could be created by using only translation.

b. Trace a smaller basic design element from which the rest of the pattern could be created by using reflection and translation. Indicate the line of symmetry and the length and direction of the translation.

c. In part b, the movement required to generate the pattern from the basic design element is called a *glide reflection*. What is the difference between a reflection and a glide reflection?

18b. OHIO

18c.
H	O	U	I
A	H	T	O
W	I	A	W
A	O	H	A
I			
I			

18d. Possible answers: COD, KICK, BOX, HIKE

18e. Possible answers:
Y	W	T	H	W
O	H	O	A	A
Y	A	M	Y	X
O	M	M		Y
	Y			

18f. See below left.

18g. Possible answers: SOS, NOON

19a.

19b. See below left.

19c. A glide reflection is a reflection followed by a translation (or a translation followed by a reflection). A reflection does not involve a translation.

18f. The angles of rotation are all 180°.

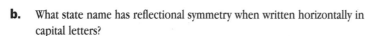

19b.

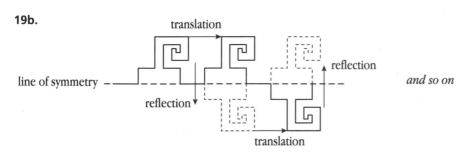

20. Below is one possible design element. There are vertical lines of symmetry through the center of each design element and through the center of each small triangle. There is no rotational symmetry.

21. Below is one possible design element. There are vertical lines of symmetry through the centers of and between design elements; a horizontal line of symmetry through the center of the design; and rotational symmetry with an angle of rotation of 180° about the center of each design element and the point between design elements.

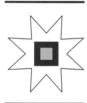

22. See right.

23. See page 23n.

In 20–23, a *strip pattern* is given. Many traditional cultures use strip patterns to decorate pottery and weavings. The patterns below are reproduced on Labsheet 1.ACE4. For each pattern, identify a basic design element, and describe the lines of symmetry and the centers and angles of rotation in the strip pattern. Imagine that each design continues forever in both directions.

20.

21.

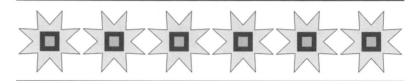

22.

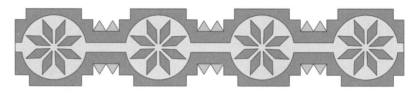

23.

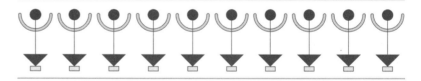

22. Below is one possible design element. There are vertical lines of symmetry through the centers of and between design elements, and a horizontal line of symmetry through the center of the design. There is rotational symmetry with an angle of rotation of 180° about the center of each design element and the point between design elements.

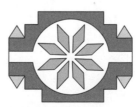

24. Look back at your answers to questions 21–23. What are the limitations on the possible angles of rotation for a strip pattern?

25. A regular hexagon can be enclosed by a circle.

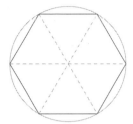

a. What are the angle measures of the triangles formed by the diagonals of the hexagon?

b. How are the side lengths of the hexagon related to the radius of the enclosing circle?

c. How are your answers to parts a and b related to the steps required to make a kaleidoscope design from a basic design element?

d. How are your answers to parts a and b related to the symmetries that occur in any kaleidoscope pattern?

26. Copy the design below. Then use tracing paper to help you sketch a full kaleidoscope design from this basic design element.

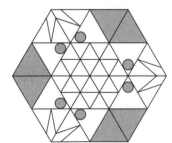
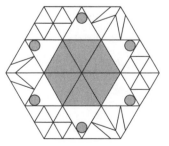

24. The only possible angle of rotation for a strip pattern is 180°. Any other rotation would result in a different orientation for the entire strip (these designs, for example, would no longer run from left to right).

25a. All three angles of each triangle measure 60°.

25b. The radius of the circle is equal to the side length of the hexagon.

25c. If the pattern within a triangle has a line of symmetry through the center of the circle to the midpoint of the opposite side, rotate the pattern piece through successive turns of 60° to complete the kaleidoscope design. The fact that each angle of the pattern piece is 60° and a complete rotation about the center is 360° means the successive turns are 60°. If the triangle does not have reflectional symmetry as described above, you must reflect this basic design triangle over a side of the triangle and then rotate the resulting two-triangle design by intervals of 120°.

25d. See page 23n.

26. There are three possible kaleidoscope designs, depending on which vertex is chosen for the center of the design.

Extensions

27. Use a compass to draw a circle with a specific radius. Do not change the opening of the compass. Place the stationary point of the compass at a point on the circle and strike an arc on the circle. Then place the point of the compass at the point where the arc intersects the circle, and strike another arc. Continue this process to mark off six points on the circle. Connect successive points to form a hexagon.

28. Possible technique: Connect every other vertex of the hexagon to make the star, and then shade the area around the star.

29. Possible technique: Connect every vertex to every other vertex. Shade every other region inside the star as shown, and shade the region outside of the hexagon.

30. Possible technique: Connect every other vertex of the hexagon to create the inscribed triangle, and shade as shown.

31. Possible technique: Connect every other vertex (two sets) of the hexagon to make a star, and then connect opposite points of the star to find the points from which to start the smaller, white star. Draw the smaller star in a similar way. Shade as shown.

Extensions

27. How could you use the information about hexagons and enclosing circles in ACE question 25 to draw a regular hexagon by using only a straightedge and compass?

In 28–33, the given design can be re-created by outlining and shading parts of a regular hexagon drawn inside of a circle. Use a compass and a straightedge to re-create each design.

28.

29.

30.

31.

32.

33.

32. Possible technique: Connect every vertex to every other vertex to create the larger star. Draw the smaller star in a similar way and shade as shown.

33. Possible technique: Connect every other vertex (two sets) and shade to create the six triangular pieces.

Mathematical Reflections

In this investigation, you looked at designs that have symmetry, and you explored some of the ways a basic design element can be used to generate a pattern with symmetry. These questions will help you summarize what you have learned:

1 **a.** How can you determine whether a design has reflectional symmetry?

b. How can you determine whether a design has rotational symmetry? If a design has rotational symmetry, how can you find the angle of rotation?

c. How can you determine whether a design has translational symmetry? If a design has translational symmetry, how can you describe the translation used to create the design?

2 **a.** Describe the tools and the process you would use to create a design with reflectional symmetry from a basic design element.

b. Describe the tools and the process you would use to create a design with rotational symmetry from a basic design element.

c. Describe the tools and the process you would use to create a design with translational symmetry from a basic design element.

Think about your answers to these questions, discuss your ideas with other students and your teacher, and then write a summary of your findings in your journal.

Possible Answers

See page 23o.

Tips for the Linguistically Diverse Classroom

Original Rebus The Original Rebus technique is described in detail in *Getting to Know Connected Mathematics*. Students make a copy of the text before it is discussed. During the discussion, they generate their own rebuses for words they do not understand; the words are made comprehensible through pictures, objects, or demonstrations. Example: Question 1b—Key words and phrases for which students might make rebuses are *rotational symmetry* (curved arrow), *angle* ($\angle$).

TEACHING THE INVESTIGATION

1.1 • Reflectional Symmetry

This problem introduces students to reflectional symmetry and the idea of a line of symmetry. Students learn ways to check for reflectional symmetry and look for examples of it in several designs.

Launch

Review the introduction to reflectional, rotational, and translational symmetry that is presented in the student edition. Then begin a discussion with students about how they might determine whether reflectional symmetry exists in a given design.

> What does the word *symmetry* mean to you? What does the word *reflection* make you think about?

Display an object or a design that has reflectional symmetry, such as the butterfly pictured in the student edition.

> Does this object have symmetry?
>
> How might you check for reflectional symmetry in an object or a design? What tools might be useful?

Recognizing the fact that the two halves of a design on either side of a line of symmetry are mirror images, students often suggest using a mirror to check for reflectional symmetry. If you have mirrors or other reflecting tools available, demonstrate their use for the class.

Encourage students to think about other ways they might check for reflectional symmetry.

> What other methods or procedures can you think of for testing whether a design has reflectional symmetry?
>
> How might you use tracing paper to find reflectional symmetry?
>
> How might you use paper folding to find lines of symmetry?

Have mirrors, transparent reflection tools, tracing paper, rulers, and other materials available for students to use. Distribute Labsheet 1.1 to each student. Have students work individually and then with partners on the problem and the follow-up.

Explore

Encourage students to be accurate as they draw each line of symmetry. Make sure they understand that the problem asks about the symmetry of the *entire design*, not of individual elements of the design such as a triangle, square, cross, or diamond.

As you circulate, ask students to think about which tools or methods are the most useful or the easiest for working with the designs in this problem.

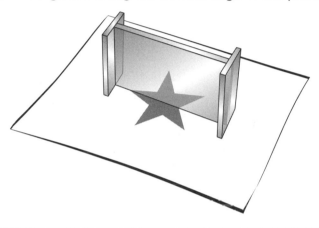
Summarize

Have students share their answers with the class, perhaps drawing each line of symmetry on Transparency 1.1 at the overhead. Encourage them to be precise as they explain their reasoning and how they used specific tools or methods to check for reflectional symmetry.

At first glance, it may appear that the design in part C has two lines of symmetry in addition to the horizontal and vertical lines through the center: those on the diagonals. However, the orientation of the eight small triangles makes these possible lines of symmetry incorrect. Such a fine detail can easily be overlooked in checking for reflectional symmetry. Explore this idea with the class.

What could we do to make this design have four lines of symmetry?

Three of many ways to make this design symmetrical about its diagonals are to remove the small triangles, to make the triangles into diamonds, or to rotate the triangles.

The design in part D has no lines of symmetry. Ask students whether the figure would demonstrate reflectional symmetry if the feet, head, and arms were removed. The answer is still no because the patterns in the upper-right and upper-left corners are not mirror images.

Review with the class the various tools they used in this problem.

> How can you use tracing paper to test a design for reflectional symmetry? *(Trace the entire design, and look for a way to fold the tracing so that one half folds exactly onto the other half. The fold line is the line of symmetry. Or, imagine where the line of symmetry is, trace the half of the design on one side of that line, and then flip the traced half over the line to see whether it matches the other half.)*

If students have used transparent reflection tools, ask how they can be used to test a design for reflectional symmetry. Students should be able to explain that they set the tool on what appears to be the line of symmetry and then observe whether the two halves of the figure match.

The quilt designs in the follow-up should yield a nice discussion about shapes that do not have reflectional symmetry and how that can be determined. The two quilt designs *do* have rotational symmetry; some students may mention the fact that the designs can be rotated to match the original orientations. You may want to discuss this idea, but make sure everyone understands that the designs do not have reflectional symmetry. You may also want to ask how the designs could be altered so that they *do* have lines of symmetry.

1.2 • Rotational Symmetry

In this problem, students examine illustrations of automobile hubcaps as they learn about the properties of rotational symmetry and the concepts of center and angle of rotation.

Launch

Have students look at the picture of the fan on page 5. Discuss the fact that although the fan does not have reflectional symmetry (students can verify this using tracing paper or transparent reflection tools), it does have another type of symmetry.

> Objects that rotate about a centerpoint are often designed so that after a partial turn, they look the same as they did in the original position. We say that such objects have *rotational symmetry*.

> What tools might be useful for checking whether the fan blades have rotational symmetry?

Some students might suggest sketching the fan on tracing paper or a blank transparency and then rotating the sketch about the center until the sketch again coincides with the original.

Discuss the pinwheel design shown in this problem; it is reproduced on Transparency 1.2A.

There are many angles through which we can rotate this pinwheel about its centerpoint so that it looks the same as it does in its original position. The smallest angle through which it can be turned to coincide with the original is called the *angle of rotation.*

Think about how you might determine the measure of the angle of rotation for designs like this pinwheel.

Have mirrors, tracing paper, rulers, angle rulers or protractors, and other tools available. Distribute Labsheet 1.2 to each student. Have students work individually on the problem and then check their ideas with their partners.

Explore

Students can measure the angle of rotation for each hubcap in any number of ways. For example, the 120° angle of rotation in the first hubcap might be measured by locating the centerpoint and then drawing lines from the centerpoint to corresponding points of two adjacent parts of the pattern as shown.

The basic idea is to determine a "sector," or section, of the hubcap that could be rotated to make the entire hubcap design. The angle of the section is the angle of rotation.

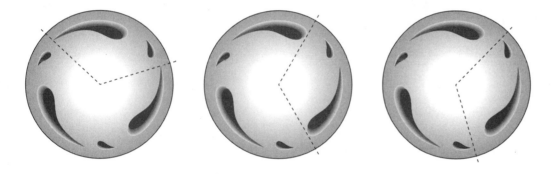

If students find determining the angles of rotation for the hubcaps easy, ask:

Can you think of a way to determine the angle of rotation *without* using a measuring tool?

This idea will be explored in the summary.

Have students who finish early work on the follow-up questions, which ask them to create their own hubcap designs. Otherwise, assign the follow-up as homework.

Summarize

Ask students to share the methods they used to determine the angle of rotation for each hubcap. Some will talk about drawing lines from the center of the hubcap to corresponding points of two adjacent parts of the pattern, thus creating an angle that can be measured with an angle ruler or a protractor. Help students to understand that each design has a particular angle of rotation regardless of which two adjacent parts are chosen.

Now discuss the reflectional symmetries in the hubcaps, going over the lines of symmetry in each. If some students have created their own hubcap designs, allow them to share the designs with the class and explain how they meet the given criteria.

Next, lead the class in an exploration of the relationship between the angle of rotation and the number of ways a particular design can be oriented to coincide with the original (or, the number of rotation angles).

> Is there a way to determine the angle of rotation for a particular design without actually measuring it?

Visualizing the smallest section of the hubcap that can be rotated to complete the hubcap is a way to determine the angle of rotation without measuring. In the design below, the indicated section would be replicated five times to complete the design. Therefore, the angle of rotation is $360° \div 5 = 72°$.

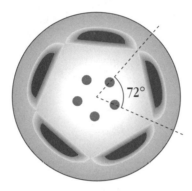

If no ideas are offered, point out the patterns in the four examples from Problem 1.2.

> In one full rotation of hubcap 1, how many times could you stop and have the hubcap look the same as it does now? *(5 times)* What is the angle of rotation for this hubcap? *(72°)* What's the measure of a full turn? *(360°)*

> In one full rotation of hubcap 2, how many times could you stop and have the hubcap look the same as it does now? *(3 times)* What is the angle of rotation for this hubcap? *(120°)* What's the measure of a full turn? *(360°)*

> What relationship do you see between the angle of rotation, the number of ways you could match the original design in a full rotation, and the measure of a full turn? *(The angle of rotation is the measure of a full turn divided by the number of ways to match the design: 360° ÷ 3 = 120° and 360° ÷ 5 = 72°.)*

Once students understand this relationship, have them apply it.

> Suppose you know the angle of rotation of a particular design. How can you use it to find all the other angles through which the design can be rotated to match the original design?

You may want to use an example, such as the star shown below, to demonstrate this idea. The star has an angle of rotation of $360° \div 6 = 60°$, so the possible angles through which it can be rotated to coincide with the original design are $1 \times 60°$, $2 \times 60°$, $3 \times 60°$, $4 \times 60°$, $5 \times 60°$, and $6 \times 60°$; in other words, $60°$, $120°$, $180°$, $240°$, $300°$, and the full rotation of $360°$.

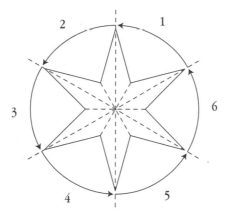

1.3 • Symmetry in Kaleidoscope Designs

In this problem, students look for reflectional and rotational symmetry in various kaleidoscope designs and are introduced to the idea of a basic design element.

Launch

Direct students' attention to the six kaleidoscope designs pictured in the student edition and reproduced on Transparency 1.3. (Note: Design 3 cannot actually be produced by a kaleidoscope because hexagonal kaleidoscope designs are formed by mirrors set at a 60° angle and contain three pairs of adjacent triangles that are mirror images. Students may discover this fact for themselves in the course of the problem.)

What basic polygon shape do these designs have? *(a hexagon)*

Do you see any reflectional symmetry in these designs? *(Most of the designs have reflectional symmetry.)*

Do you see any rotational symmetry in these kaleidoscope designs? *(All the designs show rotational symmetry.)*

How is this related to the rotational symmetries of a hexagon? *(Each rotational symmetry of a hexagon is a rotational symmetry of the kaleidoscope design.)*

Have students work in pairs to investigate the six kaleidoscope designs and answer the follow-up questions. Distribute a copy of Labsheet 1.3 to each student. Have mirrors, transparent reflecting tools, tracing paper, rulers, angle rulers or protractors, and other tools available for students to use.

Explore

Students are asked to look for reflectional and rotational symmetry in each design. If they have difficulty identifying all the symmetries in the designs, encourage them to use the various tools and techniques they have experimented with in the previous problems.

Summarize

Have students share the reflectional and rotational symmetries they found in each design. Then help them to make some general observations about the kaleidoscope designs.

> Do all the kaleidoscope designs have reflectional symmetry? *(No, design 3 does not.)*
>
> What do you notice about the angles of rotation for these kaleidoscope designs? *(They are either 60° or 120°.)*
>
> How are designs 1 and 2 similar to and different from designs 4 and 5? *(All four designs have six triangular parts. Designs 1 and 2 have 120° angles of rotation; designs 4 and 5 have 60° angles of rotation.)*
>
> Why do you think this is so? *(The designs are based on hexagons, and hexagons have 60° angles of rotation. In designs 4 and 5, the six triangular sections are all alike, so the designs have 60° angles of rotation. Designs 1 and 2, however, each have two different triangular sections. These designs must be rotated through two triangular sections, or 120°, to match with the original design.)*

Follow-up question 1 will help students discover an important connection between reflectional symmetry and rotational symmetry by leading them to uncover the mathematical relationship between them.

> How many lines of symmetry do designs 4 and 5 have? *(6)* How many lines of symmetry do designs 1, 2, and 6 have? *(3)*
>
> Why do they have different numbers of lines of symmetry? *(Each of the designs has the three lines of symmetry that pass through opposite vertices of the hexagon. In designs 4 and 5, each triangular section has a line of symmetry, so these designs have three additional lines of symmetry—those that pass through the midpoints of opposite sides of the hexagon.)*
>
> What mathematical relationship do you see between the number of lines of symmetry and the angle of rotation? *(The product of these two numbers is always 360°.)*

Have students share the various basic design elements they found for the designs.

For the Teacher: Reflecting Patterns in a Hinged Mirror

Using two mirrors, you can demonstrate the patterns created by reflections in a hinged mirror. You might allow students to experiment with the mirrors during this investigation.

1. Place the two mirrors at an angle facing each other. Tape the mirrors together so that they can be opened as if hinged.

2. Draw a dark line on a piece of paper.

3. Stand the mirrors on the line, positioned so that an equilateral triangle is reflected in the mirrors with the part of the line between the mirrors as one side of the triangle.

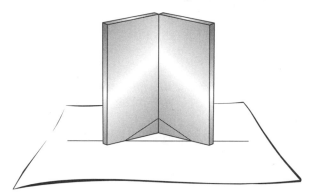

4. Reposition the mirrors by opening or closing the hinge to show a square, a pentagon, and so on.

5. Look for the lines of symmetry in each reflected image.

6. Place an object or draw a design between the line on the paper and the two faces of the mirrors; for example:

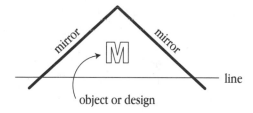

Now set the mirrors to show an equilateral triangle, a square, a pentagon, and so on, and observe the reflected images.

It is a nice extension to measure the angle between the mirrors for each polygon, make a table of the data, and generalize the angle needed to make an n-sided polygon by observing the pattern in the table. From the data below, it is evident that the angle of the mirrors necessary to produce an n-sided polygon is $\frac{360°}{n}$.

Number of sides	Angle of mirrors
3	120°
4	90°
5	72°
6	60°

In this problem, students examine several tessellations for examples of translational symmetry, and they explore connections among the three kinds of symmetry they have studied.

Launch

Ask students to point out or to cite examples of designs that exhibit translational symmetry, such as those often seen in the patterns on fabric, wallpaper, and gift wrap.

Explain that translations are also called *slides* because a pattern with translational symmetry can be created by repeatedly sliding and copying the basic design element in the pattern. You may want to illustrate this process with the idea of making a stamp of a basic design element and using that stamp repeatedly to create a pattern with translational symmetry. You might even demonstrate with a real stamp.

> How might you explain to someone how to re-create the wallpaper pattern shown on page 12?

Students may informally describe the motion, for example, as "move to the right and make a copy of the original piece."

> Would your directions be precise enough for someone to re-create this pattern exactly?

Explain that an arrow is one way to communicate exact direction and distance. Direct students' attention to the illustration of such an arrow in the student edition.

> By drawing an arrow the same length as the translation and in the desired direction, you can specify exactly the translational symmetry in a design.

Distribute Labsheet 1.4 to each student. Have students work on the problem in groups of three or four.

Explore

Each student should produce his or her own results and then compare them with the others produced by the group. When the group has finished the problem, the students should move on to the follow-up questions.

Advise students to be careful when outlining what they think is the basic design element and specifying a translation that could be used with that element to help create the tessellation. If some students are outlining an element that isn't a basic design element, you might ask:

> Is there anything simpler that would work? Could you simplify your design element?

Some students may find it easier to trace the basic design elements than to outline them. Also, you may want to distribute blank transparencies so that each group can outline what they have found to be the basic design elements for sharing in the summary.

There are various ways to outline each basic design element and to specify the translation of each basic design element. Ask students to think about whether there are other ways they could specify each basic design element, drawing their attention to the ways that others in their groups approached the problem.

In tessellation 3, some students might identify a single, shaded "arrowhead" shape as the basic design element. If so, remind them that a tessellation cannot have gaps; the white shapes must be accounted for as well.

In tessellation 4, students might ignore the coloring and choose a single bird as the basic design element. If students consider color as part of the pattern, a larger basic design element would have to be identified—one that contains a bird of each color.

Summarize

In the summary, encourage students to share the various ways that they identified the basic design element for each tessellation and specified the translation for each basic design element. It is extremely powerful for them to see that there are many correct ways to specify design elements and translations for each tessellation.

Call on a group to show a basic design element at the overhead. Ask:

> Do you all agree that this basic design element can be used to reproduce the design? Why or why not?

Once the class accepts or corrects the design element, ask:

> Let's collect every possible way you can think of to specify translations that will re-create this tessellation.

Collect as many examples as the class offers in a few minutes.

> Do you think these are the only directions we could give that would help someone re-create this tessellation? *(No, there are many more.)*

> In each different translation, is the length of the arrow the same? *(No; different directions require arrows of different lengths.)*

Students were asked to draw an arrow showing how the basic design element can be translated to produce another part of the pattern. You may want to explore with the class how many arrows would actually be needed to specify how to re-create the entire tessellation. For example:

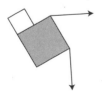

Discuss the other types of symmetries in each tessellation. To help students make a real-world connection, ask:

> If you were designing a machine to stamp out a tessellation pattern, what have you learned that would help you make the operation cost effective or require a minimum amount of time?

The discussion should center around the ideas of how much of the pattern should be stamped out at a time and how the machine or the material should move after each cycle.

Additional Answers

Answers to Problem 1.2 Follow-Up

1. Possible design:

2. Possible design:

3. Possible answer: Rotational symmetry helps some items function better—for example, fans are designed with rotational symmetry to help move air more efficiently. Rotational symmetry is also used to keep rotating objects, like propellers, well balanced. And some designs incorporate rotational symmetry for artistic effects.

Answers to Problem 1.3

A.

design 1   design 2   design 3

design 4   design 5   design 6

Answers to Problem 1.3 Follow-Up

2.

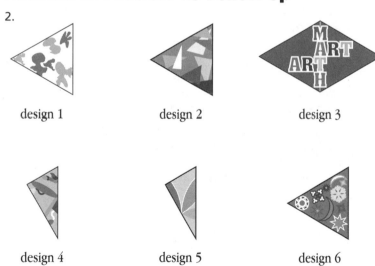

design 1   design 2   design 3

design 4   design 5   design 6

ACE Answers

Applications

8a. The design has reflectional symmetry, rotational symmetry, and translational symmetry. *Reflectional symmetry:* Since the design extends forever in all directions, there is an infinite number of lines of symmetry; six lines of symmetry are drawn below. *Rotational symmetry:* Since the design extends forever in all directions, there is an infinite number of centers of rotation. Three centers of rotation are indicated below. The design has a 60° angle of rotation about points *A* and *C,* and a 180° angle of rotation about point *B.* All other centers of rotation have the same positions relative to the basic design element as these points. *Translational symmetry:* An infinite number of translations can be specified; one is indicated by the arrow.

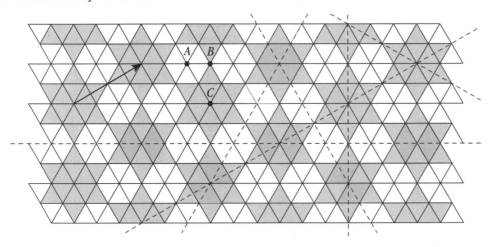

8b. To check a reflectional symmetry, you could fold the design on a possible line of symmetry to check that corresponding parts of the design match, you could place a mirror on the line to see whether the reflection looks like the design behind the mirror, or you could trace part of the pattern on one side of the line and then flip the tracing over the line to see whether the tracing matches the pattern on the other side. To check a rotational symmetry, you could trace part of the pattern and then rotate the tracing around the possible center of rotation. To check a translational symmetry, you could trace part of the pattern and then slide it along the arrow.

10c. There are diagonal lines of symmetry through the centers of the star-shaped pieces. There is 180° rotational symmetry about the center of any star-shaped piece and any diamond. There are many translational symmetries, including a vertical slide, a horizontal slide, and a diagonal slide. Some lines of symmetry, centers of rotation, and translations are indicated below.

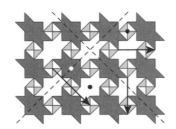

11c. There is a 90° angle of rotation about the centers of the pinwheels and a 180° angle of rotation about the center of any cluster of four white triangles. There are many translational symmetries including a vertical slide, a horizontal slide, and a diagonal slide. Some centers of rotation and translations are indicated below.

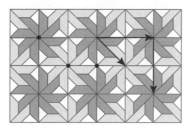

12c. There are three types of lines of symmetry, two diagonal and one vertical; each goes through the center of a hexagon and the center of an adjacent three-armed piece. There is a 120° angle of rotation about the center of any hexagon, three-armed piece, or nonagon. There are many translational symmetries including a vertical slide, a horizontal slide, and a number of diagonal slides. Some lines of symmetry, centers of rotation, and translations are indicated below.

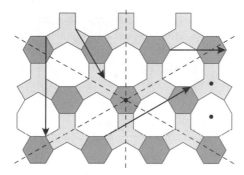

Connections

23. Below is one possible design element. There are vertical lines of symmetry through the middle and at both ends of each design element. There is no rotational symmetry.

25d. Because successive turns of 60° or 120° are made to construct it, a kaleidoscope design has rotational symmetries of either 60°, 120°, 180°, 240°, 300°, and 360° *or* 120°, 240°, and 360°. Because adjacent triangles share a common side and are mirror images, there will be a minimum of three lines of symmetry. If the basic element has an additional symmetry, the design has six lines of symmetry.

Mathematical Reflections

1a. There are several ways to determine whether a design has reflectional symmetry. For each method, you must first identify a line you think might be a line of symmetry. You can place a transparent reflection tool on the line. If the reflected image matches the image behind the tool, the design has reflectional symmetry. If the design is on a sheet of paper, you can fold the design on the line. If the two halves match, the design has reflectional symmetry. You can also trace the entire design or the part of the design on one side of the line and then flip your tracing over the line. If the tracing matches the original design, the design has reflectional symmetry.

1b. To determine whether a design has rotational symmetry, first identify the point you think might be the center of rotation. Then trace the entire design, or just the basic design element, and rotate the tracing around the point. If the tracing matches the original design at some point before you finish a complete rotation, the design has rotational symmetry. You can find the angle of rotation by finding corresponding points on two consecutive basic design elements and then measuring the angle formed by line segments from the center of rotation to those points or by determining the total number of ways the design could be rotated to match with the original and dividing 360° by that number.

1c. To determine whether a design has translational symmetry, trace the entire design or the part you think might be a basic design element. If you can slide your tracing in a straight line so that it matches with the original design, the design has translational symmetry. You can describe a translation by drawing an arrow showing the length and direction of the translation or by giving the length and direction in words.

2a. You can use a transparent reflection tool or tracing paper to create a design with reflectional symmetry. First, you need to draw the basic design element and a line of reflection. To use the reflection tool, place the edge of the tool on the line of reflection. Look at the reflection of the basic design element, and trace the image on the other side of the line. To use tracing paper, trace the basic design element and the line of reflection on the tracing paper and then flip the tracing over the line, matching the line of reflection on your tracing with the line of reflection on the original. Press on key points on the tracing to make indentations on the final paper. Then remove the tracing paper and connect the indentations to create the final image. Instead of making indentations, you can retrace the figure through the back of the tracing paper. This will leave light pencil marks that you can retrace to show the image more clearly.

2b. To create a design with rotational symmetry, draw the basic design element and a center of rotation. Trace the design element and the center of rotation, and then rotate the tracing through the desired angle about the center, using an angle ruler or a protractor to measure the angle from a key point on the original element to the corresponding point on the tracing. (The angle of rotation must be a factor of 360°.) Copy the rotated tracing onto the final paper. Repeat this process until you have made a complete rotation about the center of rotation. If you have a stamp of the basic design element, you can rotate the stamp through the desired angle of rotation and stamp the image.

2c. To create a design with translational symmetry, you can use tracing paper and a ruler. First, draw a basic design element and trace it onto tracing paper. Then use the ruler to help you slide the tracing the desired direction and distance, and copy the tracing onto the final paper. Repeat this process several times. If you have a stamp of a basic design element, you can slide the stamp the desired direction and distance and stamp the image.

Symmetry Transformations

This investigation introduces the mathematical language for describing rigid motions of geometric figures. Building on their hands-on experiences, students formulate precise mathematical descriptions for performing transformations.

In the first three problems—Problem 2.1, Describing Line Reflections; Problem 2.2, Describing Translations; and Problem 2.3, Describing Rotations—students observe properties that define each transformation and would allow someone else to perform it. While each transformation is associated with a type of symmetry, not every transformation produces a symmetric design. Line reflections match points of a figure to points of an image figure, and the original figure and its image form a symmetric figure. The rotation of a figure will produce a symmetric design if the angle of rotation is a factor, k, of 360 and the design has $360 \div k$ figures. And, if a translation of a figure is repeated infinitely, a strip design with translational symmetry is produced.

In Problem 2.4, Combining Transformations, students investigate which single transformation is equivalent to a given sequence of two transformations.

Mathematical and Problem-Solving Goals

■ **To examine reflections, translations, and rotations to determine how to specify such transformations precisely**

■ **To use the properties of reflections, translations, and rotations to perform transformations**

■ **To find lines of reflection, magnitudes and directions of translations, and centers and angles of rotation**

■ **To examine the results of combining reflections over two intersecting lines or two parallel lines; two translations; or two rotations to find single a transformation that will produce the same result**

	Materials	
Problem	**For students**	**For the teacher**
All	Graphing calculators, tracing paper (or lightweight plain paper), rulers, angle rulers or protractors, compasses, mirrors, transparent reflection tools such as Image Reflectors	Transparencies: 2.1A to 2.4B, overhead graphing calculator, transparent ruler, chalkboard compass (all optional)
2.1	Labsheets 2.1A and 2.1B (1 each per student)	Transparency of Labsheet 2.1B (optional)
2.2	Labsheet 2.2 (1 per student)	Transparency of Labsheet 2.2 (optional)
2.3	Labsheets 2.3A and 2.3B (1 each per student), blank transparencies	Transparency of Labsheet 2.3B (optional), blank transparencies
2.4	Labsheets 2.4A and 2.4B (1 each per student)	Transparency of Labsheet 2.4B (optional)
ACE	Labsheets 2.ACE1, 2.ACE2, 2.ACE3, and 2.ACE4 (1 each per student); tracing paper	Transparencies of Labsheets 2.ACE1, 2.ACE2, 2.ACE3, and 2.ACE4 (optional)

Launch

- Introduce the ideas of transformations and images.

- Have groups of two to four work on the problem.

Explore

- Ask students to look for patterns in the measurements they are making.

Summarize

- Have students share what they have learned about line reflections.

- Introduce the term *perpendicular bisector.*

- Assign and then review the follow-up.

INVESTIGATION 2

Symmetry Transformations

You can draw a figure with reflectional symmetry by finding the mirror image of polygon *ABCDE* over the line. To help you draw your figure, you could set a mirror on the line to see what the image looks like.

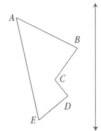
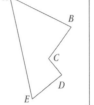

Reflect polygon *ABCDE* over the line.   Polygon *A′B′C′D′E′* is the image of polygon *ABCDE*.

Drawing the mirror image of a figure is an example of a geometric operation called a *transformation*. **Transformations** produce a copy of a figure in a new position. The copy is called the *image* of the original figure.

In this investigation, you will explore transformations associated with different types of symmetry, and you will learn to give precise mathematical instructions for performing transformations.

2.1 Describing Line Reflections

Transformations that create figures with reflectional symmetry are called **line reflections**. When you make a design using a line reflection or any other transformation, it helps to know precisely how the transformation matches each point on a figure to a point on the image of the figure. In this problem, you will consider this question:

What is the relationship between a figure and its image under a line reflection?

ACE questions 1–4, 9, 19, and unassigned choices from earlier problems

Answers to Problem 2.1

A. Each vertex and its image are the same distance from the line of reflection. The segment joining each vertex to its image is perpendicular to, or forms a right angle with, the line of reflection.

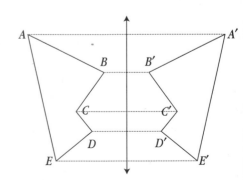

Problem 2.1

A. In the figure on page 24, polygon $A'B'C'D'E'$ is the image of polygon $ABCDE$ under a line reflection. This figure is reproduced on Labsheet 2.1A.

    1. Draw a line segment from each vertex of polygon $ABCDE$ to its image on polygon $A'B'C'D'E'$.

    2. Measure the angles formed by each segment you drew and the line of reflection.

    3. For each vertex of polygon $ABCDE$, measure the distance from the vertex to the line of reflection along the segment you drew and the distance from the line of reflection to the image of the vertex.

    4. Describe the patterns in your measurements from parts 2 and 3.

B. The figure at right is reproduced on Labsheet 2.1A. Use what you discovered in part A to draw the image of polygon $JKLMN$ under a reflection over the line. Use only a pencil, a ruler, and an angle ruler or protractor. Describe how you drew the image.

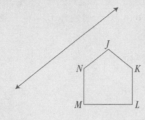

C. The reflection-symmetric design below is reproduced on Labsheet 2.1A. Use only a pencil, a ruler, and an angle ruler or protractor to find the line of symmetry for this design. Describe how you found the line of symmetry.

D. Complete this definition of a line reflection: A *line reflection* matches each point X on a figure to an image point X' so that . . .

B. Draw a line from each vertex on polygon *JKLMN* to the line of reflection and extending beyond it. Measure the length of the line segment from each vertex to the line of reflection, and measure an equal distance from the line of reflection to the opposite side. Mark the new points *J'*, *K'*, *L'*, *M'*, and *N'*, and connect them.

C. See page 41p.

D. A *line reflection* matches each point *X* on a figure to an image point *X'* so that (1) the distance from each of these paired points to the line of symmetry is the same and (2) the segment *XX'* is perpendicular to the line of symmetry.

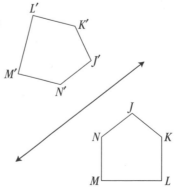

■ **Problem 2.1 Follow-Up**

1. a. Draw a shape and a line. Then draw the image of your shape under a reflection over the line.

b. What information would you need to give a classmate so that, starting with your original drawing, he or she could duplicate your line reflection?

c. Explain how your shape and its image fit the definition of *line reflection* you wrote in part D of Problem 2.1.

2. The figures below appear on Labsheet 2.1B. Notice that in two of the figures, the shape crosses the line of reflection. Do parts a and b for each figure.

a. Draw the image of the shape under a reflection over the line.

b. Does the final drawing have reflectional symmetry? Explain.

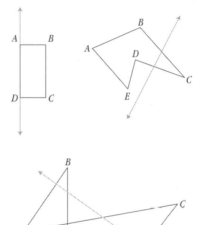

3. What tools are useful for performing a line reflection? Carefully describe how each tool is used.

Answers to Problem 2.1 Follow-Up

1. a. Drawings will vary.

   b. For someone else to duplicate the line reflection, all that must be specified is the line of reflection.

   c. Answers will vary. Students should explain how their example illustrates their definition.

2. See page 41p.

The figures you have reflected in this unit are two-dimensional. Two-dimensional figures are drawn in a *plane*. You can picture a plane as a sheet of paper that extends infinitely in all directions. When you reflect a figure over a line, you can imagine that you are reflecting the entire plane over the line. The figure just goes along for the ride.

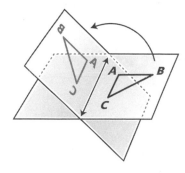

4. When you perform a line reflection, are any points in the plane in the same location *after* the reflection? Explain your answer.

Points that are in the same location after a transformation are called *fixed points*. Some transformations have fixed points; other transformations move every point in the plane to a new location. As you consider the transformations in the rest of this investigation, you will be asked to think about fixed points.

2.2 Describing Translations

You have seen how line reflections are related to reflectional symmetry. Other types of symmetry also have associated transformations. In Investigation 1, you saw how patterns with translational symmetry could be created by translating, or sliding, a basic pattern. A translation is a type of transformation. In this problem, you will look at examples of translations in search of an answer to this question:

What is the relationship between a figure and its image under a translation?

As you work on each example, think about the instructions you could give so that someone else could re-create the translation exactly.

At a Glance

Grouping:
pairs

Launch

■ Explain that the class will now look for a way to specify translations.

■ Have pairs work on the problem.

Explore

■ Ask students to do their own work and then share ideas with their partners.

Summarize

■ Talk about students' ideas for describing translations.

■ Assign and then discuss the follow-up.

---

3. Possible answer: A ruler is useful for measuring distances. An angle ruler or protractor is useful for drawing lines perpendicular to a line of reflection. A transparent reflection tool is useful for checking your work. Tracing paper is useful for performing reflections: you can trace the original and the line of reflection and flip the tracing paper over to see where the vertices fall on the other side of the line of reflection.

4. Only points that lie on the line of reflection are in the same location after the line reflection.

Assignment Choices

ACE questions 5, 6, 11, 17, and unassigned choices from earlier problems

Problem 2.2

A. Each diagram below shows polygon *ABCDE* and its image under a translation. These figures are reproduced on Labsheet 2.2. Do parts 1 and 2 for each diagram.

diagram 1

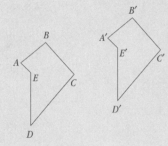

diagram 2

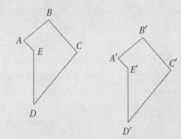

    1. Draw a line segment from each vertex of polygon *ABCDE* to its image.

    2. Describe the relationship among the line segments you drew.

B. The translations in part A slide polygon *ABCDE* onto its image, polygon *A'B'C'D'E'*. Do parts 1–3 for each diagram in part A.

    1. By performing the same translation that was used to slide polygon *ABCDE* to polygon *A'B'C'D'E'*, slide polygon *A'B'C'D'E'* to create a new image. Label the image *A"B"C"D"E"*.

    2. Polygon *A"B"C"D"E"* is the image of polygon *ABCDE* after two identical translations. How is polygon *A"B"C"D"E"* related to polygon *ABCDE*?

    3. Does your final drawing have translational symmetry? Explain.

C. Complete this definition of a translation: A *translation* matches any two points *X* and *Y* on a figure to image points *X'* and *Y'* so that . . .

Answers to Problem 2.2

A, B. See page 41p.

C.   Possible answer: A *translation* matches any two points *X* and *Y* on a figure to image points *X'* and *Y'* so that (1) the distance from *X* to *X'* is equal to the distance from *Y* to *Y'* and (2) the line *XX'* is parallel to the line *YY'*.

Answers to Problem 2.2 Follow-Up

1.   a.   Ella is indicating that each vertex of the figure should be translated in the direction of her arrow and a distance equal to the length of the arrow.

      b.

Describing Rotations

■ **Problem 2.2 Follow-Up**

1. Ella specified a translation of a polygon by drawing an arrow.

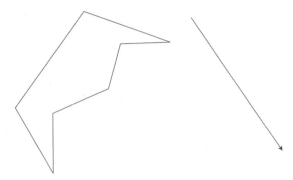

    a. How would you interpret Ella's drawing?

    b. On Labsheet 2.2, perform the translation Ella specified.

2. a. Draw a shape. Then draw the image of your shape under a translation.

    b. What information would you need to give a classmate so that, starting with your original drawing, he or she could duplicate your translation?

    c. Explain how your shape and its image fit the definition of *translation* you wrote in part C of Problem 2.2.

3. What tools are useful for performing a translation? Carefully describe how each tool is used.

4. When you translate a figure, you can imagine that you are translating the entire plane and bringing the figure along for the ride. When you perform a translation, are any points in the plane in the same location *after* the translation? That is, are there any *fixed points*? Explain your answer.

2.3 Describing Rotations

In Investigation 1, you explored rotational symmetry, and you created symmetric designs by rotating figures about a point. A rotation is a type of transformation. The rotations in this unit are *counterclockwise* turns about a point.

In this problem, you will try to answer this question:

What is the relationship between a figure and its image under a rotation?

As you work on each example, think about the instructions you could give so that someone else could re-create the rotation exactly.

Launch

■ Introduce the idea of rotational transformation.

■ Read through the problem with the class.

■ Have pairs work on the problem.

Explore

■ Help students understand how to measure the angles.

■ Let students use tracing paper to see how the vertices move.

Summarize

■ Have students share what they learned in the problem.

■ As a class, explore the properties of points on a perpendicular bisector.

■ Assign and then review the follow-up.

---

2. a. Drawings will vary.

    b. For someone else to duplicate the translation, the distance and the direction of the translation must be specified.

    c. Answers will vary. Students should explain how their example illustrates their definition.

3. Possible answer: Tools for measuring length and angles are useful for reproducing a shape exactly and in the right place. Tracing paper is useful for performing translations and drawing the final image. First, trace the original figure and the arrow indicating the translation. Move the tracing paper by sliding your arrow along the original arrow until the end of your arrow is touching the tip of the original arrow. Mark the image's vertices through the tracing paper, remove the paper, and draw the image.

4. Because the entire plane is translated, there are no fixed points.

Assignment Choices

ACE questions 7, 8, 10, 16, 18, 20, and unassigned choices from earlier problems

Problem 2.3

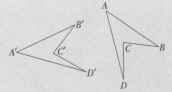

A. In the figure on the right, polygon *A'B'C'D'* is the image of polygon *ABCD* under a rotation of 60° about point *P*. This figure is reproduced on Labsheet 2.3A.

•P

1. What relationship would you expect to find between each vertex, its image, and point *P*?

2. For each vertex of polygon *ABCD*, find the measure of the angle formed by the vertex, point *P*, and the image of the vertex. For example, find the measure of angle *APA'*.

3. For each vertex of polygon *ABCD*, find the distance from the vertex to point *P* and the distance from the image of the vertex to point *P*. For example, find *AP* and *A'P*.

4. What patterns do you see in your measurements? Do these patterns confirm the conjecture you made in part 1?

B. The figures below are reproduced on Labsheet 2.3A. Do parts 1 and 2 for each figure.

1. Perform the indicated rotation, and label the image vertices appropriately.

2. Describe the path each vertex follows under the rotation.

Rotate 90° about point *P*.

Rotate 45° about point *Q*.

•P

C. For the figures in part B, use the specified rotation to rotate the *image* of the original polygon. The result is the image of the original polygon after two identical rotations. How does the location of the final image compare with the location of the original polygon? Be very specific.

D. Complete this definition of a rotation: A *rotation* of *d* degrees about a point *P* matches any point *X* on a figure to an image point *X'* so that . . .

Answers to Problem 2.3

A. **1.** Possible answer: If you formed angles by connecting each vertex and its image to the center of rotation, they might measure 60° because the angle of rotation is 60°. Also, each vertex and its image might be the same distance from the center of rotation.

2. All the angles measure 60°.

3. Segments *AP* and *A'P* measure 3.8 cm, segments *BP* and *B'P* measure 3.4 cm, segments *CP* and *C'P* measure 2.75 cm, and segments *DP* and *D'P* measure 1.5 cm.

4. The angle formed by each vertex, point *P*, and the image of the vertex is 60°: ∠*A'PA* = ∠*B'PB* = ∠*C'PC* = ∠*D'PD* = 60°. This seems to confirm the conjecture in part 1. Also, each vertex and its image are the same distance from point *P*: *AP* = *A'P*, *BP* = *B'P*, *CP* = *C'P*, *DP* = *D'P*.

B–D. See page 41q.

■ **Problem 2.3 Follow-Up**

1. a. Draw a shape and a point. Then draw the image of your shape under a rotation about the point.

b. What information would you need to give a classmate so that, starting with your original drawing, he or she could duplicate your rotation?

c. Explain how your figure and its image fit the definition of *rotation* you wrote in part D of Problem 2.3.

2. Polygon *A′B′C′D′E′F′* is the image of polygon *ABCDEF* after a rotation. This figure is reproduced on Labsheet 2.3B.

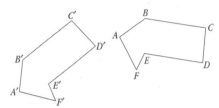

a. Find the center of rotation, and describe how you found it.

b. Find the angle of the rotation, and describe how you found it.

3. On Labsheet 2.3B, draw a line segment from a vertex of polygon *ABCDEF* to its image. The midpoint of this segment is the same distance from the original vertex as it is from the image of that vertex. Find some other points that are the same distance from the vertex and the image of the vertex. What do all of these points have in common?

4. After a polygon is rotated, how does the distance from a vertex of the polygon to the center of rotation compare to the distance from the image of the vertex to the center of rotation?

5. Look back at the rotations you have seen in this problem. Do the final drawings after the rotations have rotational symmetry? Explain.

6. What tools are useful for performing a rotation? Carefully describe how each tool is used.

7. When you rotate a figure, you can imagine that you are rotating the entire plane and bringing the figure along for the ride. When you perform a rotation, are any points in the plane in the same location after the rotation? That is, are there any *fixed points*? Explain your answer.

Answers to Problem 2.3 Follow-Up

1. a. Drawings will vary.

   b. For someone else to duplicate the rotation, the angle of rotation must be specified.

   c. Answers will vary. Students should explain how their example illustrates their definition.

2–7. See page 41r.

2.4

Combining Trans-formations

At a Glance

Grouping:
pairs

Launch

- Introduce the idea of performing two reflections over two intersecting or two parallel lines.

- Have pairs work on the problem.

Explore

- Caution students to make their drawings neatly.

- If students have trouble, suggest that they use tracing paper to find the second image.

Summarize

- Have students share what they learned about reflecting a figure over two intersecting or parallel lines.

- Assign and then review the follow-up.

2.4 Combining Transformations

Now you will explore what happens when you perform two transformations in a row, the first on the original figure and the second on the image of that figure.

Problem 2.4

A. 1. The figure below is reproduced on Labsheet 2.4A. Reflect triangle *ABC* over line 1. Then reflect the image over line 2. Label the final image *A"B"C"*.

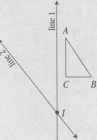

2. For each vertex of triangle *ABC*, measure the angle formed by the vertex, point *I*, and the image of the vertex. For example, measure angle *AIA"*. What do you observe?

3. For each vertex of triangle *ABC*, compare the distance from the vertex to point *I* with the distance from the image of the vertex to point *I*. What do you observe?

4. Could you move triangle *ABC* to triangle *A"B"C"* with a single transformation? If so, describe the transformation.

5. Make a conjecture about the result of reflecting a figure over two intersecting lines. Test your conjecture with an example.

B. 1. What will happen if you reflect a figure over a line and then reflect the image over a second line that is *parallel* to the first line? Would the combination of the two reflections be equivalent to a single transformation?

2. Test your conjecture from part 1 on several examples, including the one at right. Do the results support your conjecture? If so, explain why. If not, revise your conjecture to better explain your results.

Kaleidoscopes, Hubcaps, and Mirrors

Assignment Choices

ACE questions 12–15, 21–23, and unassigned choices from earlier problems

Assessment

It is appropriate to use the quiz after this problem.

Answers to Problem 2.4

A. 1.

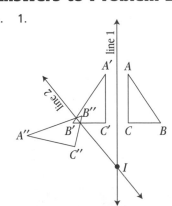

2. All the angles measure 78°.

3. The distances in each pair are equal.

4. You could rotate triangle *ABC* about point *I* through an angle of 78°.

5. Conjectures will vary. A reflection of a figure over two intersecting lines is equivalent to a single rotation with the point of intersection as the center of rotation and an angle of rotation that is twice the measure of the angle between the lines.

B. See page 41s.

■ **Problem 2.4 Follow-Up**

1. a. What single transformation is equivalent to reflecting a figure over a line and then reflecting the image over an intersecting line?

b. What single transformation is equivalent to reflecting a figure over a line and then reflecting the image over a parallel line?

2. The figure below is reproduced on Labsheet 2.4B. Draw two parallel lines so that if you reflect triangle *ABC* over one line and then reflect the image over the other line, the result will be triangle *A″B″C″*. Explain how you found the lines.

3. The figure below is reproduced on Labsheet 2.4B. Draw two intersecting lines so that if you reflect polygon *ABCD* over one line and then reflect the image over the other line, the result will be polygon *A″B″C″D″*. Explain how you found the lines.

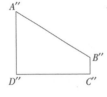
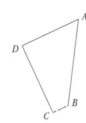

4. What would happen if you translated a figure and then translated the image? Would the combination of translations be equivalent to a single transformation? Make a conjecture about combining two translations. Test your conjecture on several examples.

5. What would happen if you rotated a figure about a point and then rotated the image about the same point? Would the combination of rotations be equivalent to a single transformation? Make a conjecture about combining two rotations. Test your conjecture on several examples.

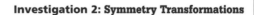

Answers to Problem 2.4 Follow-Up

1. a. Answers will vary. The single transformation that will produce the same result is a rotation about the intersection point of the lines, with an angle of rotation equal to twice the angle between the lines of reflection.

   b. Answers will vary. The single transformation that will produce the same result is a translation, or slide, perpendicular to the lines of reflection. The length of the translation is twice the distance between the lines of reflection.

2–4. See page 41s.

5. With the same center of rotation, you just add the amounts of rotations to find the amount of an equivalent, single rotation.

Answers

Applications

1. See illustration below right. Triangle *A'B'C'* is the image of triangle *ABC*. Each point on triangle *ABC* is matched to an image point on the other side of the line of reflection. The image point lies on a line passing through the original point and perpendicular to the line of reflection. The distance from the image point to the line of reflection is equal to the distance from the original point to the line of reflection.

2. See illustration below right. Polygon *A'B'C'D'E'F'* is the image of polygon *ABCDEF*. Each point on polygon *ABCDEF* is matched to an image point on the other side of the line of reflection. The image point lies on a line passing through the original point and perpendicular to the line of reflection. The distance from the image point to the line of reflection is equal to the distance from the original point to the line of reflection.

3. Possible answer:

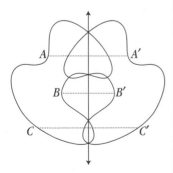

As you work on these ACE questions, use your calculator whenever you need it.

Applications

In 1 and 2, draw the image of the polygon under a reflection over the line. Describe the image of each point of the original polygon under the reflection. These figures are reproduced on Labsheet 2.ACE1.

1.

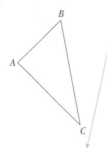

2.

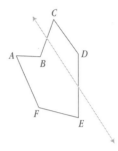

3. Shown below are a shape and its image under a line reflection. The figure is reproduced on Labsheet 2.ACE1.

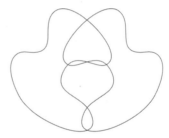

a. Draw the line of symmetry for the figure.

b. Label three points on the figure, and label the corresponding image points.

1.

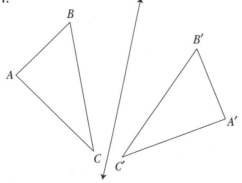

2.

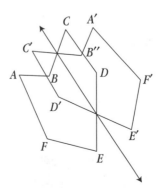

4. Shown at right are a shape and its image under a line reflection. The figure is reproduced on Labsheet 2.ACE1.

   a. Draw the line of symmetry for the figure.

   b. Label three points on the figure, and label the corresponding image point.

In 5 and 6, perform the translation indicated by the arrow. Describe the relationship between each point on the original polygon and its image point under the translation. These figures are reproduced on Labsheet 2.ACE2.

5.

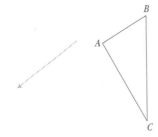

6.

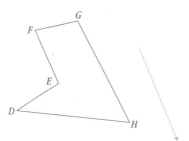

4. Possible answer:

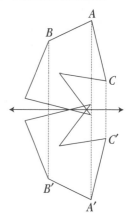

5. See illustration below left. Triangle *A'B'C'* is the image of triangle *ABC*. Each point on triangle *ABC* is matched to an image point whose distance and direction from the original point are determined by the arrow.

6. See illustration below left. Polygon *A'B'C'D'E'* is the image of polygon *ABCDE*. Each point on polygon *ABCDE* is matched to an image point whose distance and direction from the original point are determined by the arrow.

5.

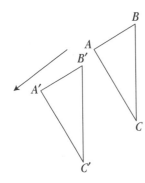

6.

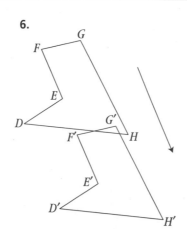

7. Triangle *A'B'C'* is the image of triangle *ABC*. Each point *X* on triangle *ABC* is matched to an image point *X'* so that *RX = RX'* and the measure of angle *XRX'* is 90°.

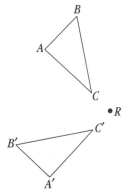

8. Polygon *A'B'C'D'E'F'* is the image of polygon *ABCDEF*. Each point *X* on polygon *A'B'C'D'E'F'* is matched to an image point *X'* so that *RX = RX'* and the measure of angle *XRX'* is 180°.

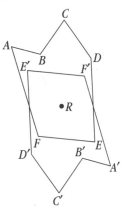

7. On Labsheet 2.ACE2, rotate triangle *ABC* 90° about point *R*. Describe the relationship between each point on the original polygon and its image point under the rotation.

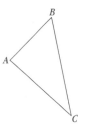

8. On Labsheet 2.ACE2, rotate polygon *ABCDEF* 180° about point *R*. Describe the relationship between each point on the original polygon and its image point under the rotation.

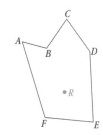

In 9–11, refer to this diagram, which is reproduced on Labsheet 2.ACE3.

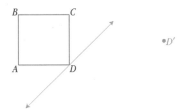

9. Draw the image of square *ABCD* under a reflection over the line.

10. Draw the image of square *ABCD* under a 45° rotation about point *A*.

11. Draw the image of square *ABCD* under the translation that slides point *D* to point *D′*.

In 12–15, a polygon and its image under a transformation are given. Decide whether the transformation was a *line reflection,* a *rotation,* or a *translation.* Then, on Labsheet 2.ACE4, indicate the line of reflection, the center and angle of rotation, or the direction and distance of translation.

12.

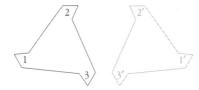

13.

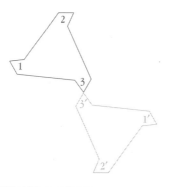

14.

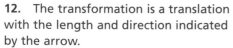

9.

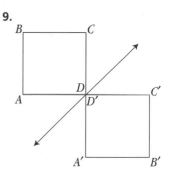

10.

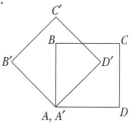

11. See page 41u.

12, 13. See below left.

14. The transformation is a 180° rotation about the point shown.

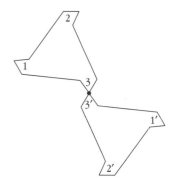

12. The transformation is a translation with the length and direction indicated by the arrow.

13. The transformation is a reflection over the line shown.

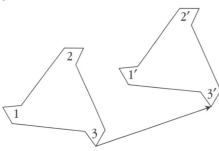

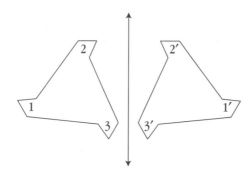

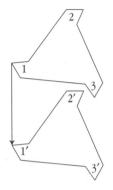

15. The transformation is a translation with the length and direction indicated by the arrow.

Connections

16. Possible answer: A square has four lines of symmetry: the vertical line through the center, the horizontal line through the center, and the two diagonals. A square also has rotational symmetry about its centerpoint with a 90° angle of rotation.

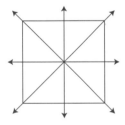

17. See page 41u.

18. Possible answer: This design has rotational symmetry about its centerpoint with a 90° angle of rotation. It does not have reflectional symmetry because there is no way to fold the design so that the halves match.

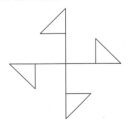

15.

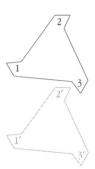

Connections

16. Create a design that has both reflectional symmetry and rotational symmetry. Explain how your design shows both types of symmetry.

17. Create a design that has both reflectional symmetry and translational symmetry. Explain how your design shows both types of symmetry.

18. Create a design that has rotational symmetry, but no reflectional symmetry. Explain how your design fits these constraints.

19. Create a design that has reflectional symmetry but no rotational symmetry. Explain how your design fits these constraints.

20. Create a design that has translational symmetry, but no reflectional symmetry or rotational symmetry. Explain how your design fits these constraints.

19. Possible answer: This design has reflectional symmetry over the line shown, but it has no rotational symmetry because there is no way to rotate it so that it looks like the original design.

20. See page 41u.

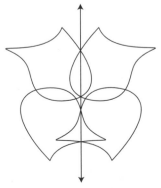

Extensions

21. Copy the figures below.

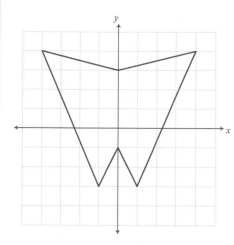

a. Determine whether each polygon has reflectional symmetry about the *y*-axis.

b. Sketch the image of each polygon under a line reflection over the *y*-axis.

c. How does the reflection image of the symmetric figure differ from the reflection image of the nonsymmetric figure?

d. Make a conjecture about whether the difference you observed in part c will occur for *any* pair of polygons in which one polygon has reflectional symmetry over the *y*-axis and the other does not. Test your conjecture on several examples, and see if you can find reasons for the patterns you observe.

21a. The left figure has reflectional symmetry about the *y*-axis; the right figure does not.

21b. See below left.

21c. The image of the symmetric figure coincides with the original figure; the image of the non-symmetric figure does not.

21d. If a figure with reflectional symmetry over the *y*-axis is reflected over that axis, the image will coincide with the original figure. If a figure that does not have reflectional symmetry over the *y*-axis is reflected over that axis, the image will not coincide with the original figure. This means that the design with reflectional symmetry looks the same after the reflection. The design without reflectional symmetry does not look the same, but the design and its image together form a symmetric design.

21b.

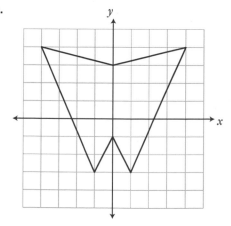

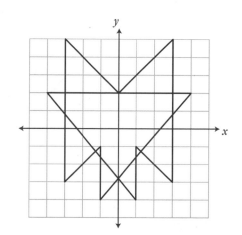

22a. The left figure has rotational symmetry about the origin; the right figure does not.

22b. See below right.

22c. The image of the symmetric figure coincides with the original figure; the image of the non-symmetric figure does not.

22d. If a figure has rotational symmetry about the origin, some rotation about the origin will result in an image that coincides with the original figure. If a figure does not have rotational symmetry about the origin, no rotation about the origin will result in an image that coincides with the original figure. For a design to have rotational symmetry, there must be pairs of vertices on the figure that are the same distance from the center. For the figure on the right, there are no points that the vertices can be paired with to make rotational symmetry possible.

23a. Possible answer: A figure has reflectional symmetry if it can be reflected over a line so that the image coincides with the original figure.

23b. Possible answer: A figure has rotational symmetry if it can be rotated less than a full turn about a point so that the image coincides with the original figure.

22. Copy the figures below.

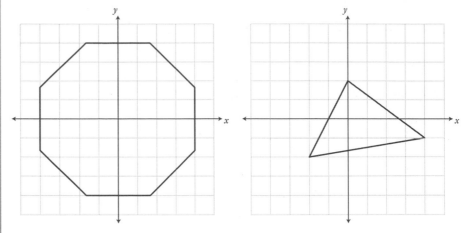

a. Determine whether each polygon has rotational symmetry about the origin.

b. Sketch the image of each polygon under a rotation of 45° about the origin.

c. How does the rotation image of the symmetric figure differ from the rotation image of the nonsymmetric figure?

d. Make a conjecture about whether the difference you observed in part c will occur for any pair of polygons in which one polygon has rotational symmetry about the origin and the other does not. Test your conjecture on several examples, and see if you can find reasons for the patterns you observe.

23. Use the language of line reflections and rotations to explain what it means for figures to have

a. reflectional symmetry.

b. rotational symmetry.

22b.

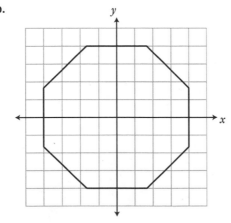

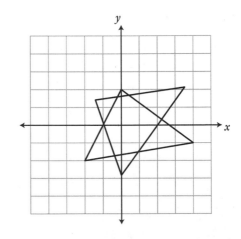

Mathematical Reflections

In this investigation, you explored line reflections, rotations, and translations—the mathematical transformations that create symmetric designs. These questions will help you summarize what you have learned:

1 Describe the relationship between a point and its image under a reflection over a line.

2 Describe the relationship between a point and its image under a rotation of *d* degrees about a point *P*.

3 Describe the relationship between a point and its image under a translation of a given length and direction.

4 What single transformation will produce the same result as reflecting a figure over a line and then reflecting the image over an intersecting line?

5 What single transformation will produce the same result as reflecting a figure over a line and then reflecting the image over a parallel line?

Think about your answers to these questions, discuss your ideas with other students and your teacher, and then write a summary of your findings in your journal.

Possible Answers

1. The image of a point lies on the line through that point and perpendicular to the line of symmetry. The distance from the original point to the line of symmetry is equal to the distance from its image to the line of symmetry.

2. The distance from a point *A* to point *P* is equal to the distance from its image, point *A'*, to point *P*. The measure of angle *APA'* is *d* degrees.

3. The image of a point under a translation is the given distance and direction from the original point.

4. A rotation about the intersection point of the lines with an angle of rotation equal to twice the angle formed by the lines will produce the same result.

5. A translation perpendicular to the lines of reflection and with a length equal to twice the distance between the lines will produce the same result.

Tips for the Linguistically Diverse Classroom

Original Rebus   The Original Rebus technique is described in detail in *Getting to Know Connected Mathematics*. Students make a copy of the text before it is discussed. During the discussion, they generate their own rebuses for words they do not understand; the words are made comprehensible through pictures, objects, or demonstrations. Example: Question 4—Key words and phrases for which students might make rebuses are *single* (1), *transformation* (curved arrow; line through the center of a symmetrical design; design with line segments connecting two points with their respective image points), *same result* (two identical images), *intersecting line* (+).

TEACHING THE INVESTIGATION

2.1 • Describing Line Reflections

In this problem, students will explore the geometric transformation associated with reflectional symmetry. They will analyze the relationship between a figure and its image under a line reflection, looking for mathematical ways to describe how such a transformation matches each point in a figure to its image.

Launch

Ask students to keep their books closed. Draw the following figure (which appears in the text) on the board or the overhead, and ask students to copy it.

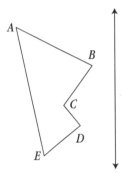

Now ask students to draw the *mirror image* of the polygon over the line, labeling matching points *A'*, *B'*, and so on.

Next, pose a question similar to the open-ended question asked at the start of Problem 2.1:

> What is the relationship between the figure you started with and the mirror image you drew?

Listen to students' ideas and how they support them. If necessary, ask about the relationship between a vertex and its image. This exercise will help students begin to think about the concepts they will encounter throughout this investigation.

> Drawing the mirror image of a figure is one kind of transformation. A *transformation* is a geometric operation that produces a copy of a figure in a new position. We call the copy the *image* of the original figure.

Read through the problem with students so that they understand what is expected of them. Mention that they will be asked to give evidence to support their answers to each part of the problem during the summary.

Have students work on the problem in groups of two to four. Save the follow-up until after the summary of the problem.

Explore

Circulate as groups work, asking questions to assess their understanding.

> What patterns do you notice in the measurements you are making?
>
> How do the measurements you are making relate to the image you would get if you used a mirror?

If students are having trouble producing the reflection image for the figure in part B, encourage them to draw from what they learned in part A.

> Look at what you discovered in part A about the relationship between a figure and its image. How does a vertex on a polygon correspond to a vertex on the image of that polygon? *(Each vertex and its image are an equal, perpendicular distance from the line of reflection.)*
>
> How can you use this relationship to draw the image of the polygon in part B?

Once students have drawn the image, you may want to suggest that they use a transparent reflection tool, a mirror, or tracing paper to check their work.

Summarize

Ask students to share what they have learned from their work with line reflections.

> When you connected vertices and images of those vertices using line segments, what did you observe about the collection of line segments? *(The line segments appear to be parallel.)*
>
> What did you observe about the line segments relative to the line of symmetry? *(The line segments appear to be perpendicular to the line of symmetry.)*
>
> What did you notice about the line segment from the line of symmetry to a vertex and the corresponding line segment from the line of symmetry to the image of that vertex? *(The line segments are of equal length.)*
>
> What can you say about the point of intersection of the line of reflection and the line segment from a vertex to its image? *(The point of intersection is the midpoint of the segment.)*
>
> Summarize what these observations tell you about how to perform a line reflection using only a ruler and an angle ruler or protractor. *(Draw a line segment from a vertex of the original figure perpendicular to the line of reflection, and measure its length. Extend the segment, and mark a point on it that is the same distance from the line of reflection but on the other side. This is the location of the image of the vertex. Do this for every vertex, and then connect the image vertices in the corresponding order to form the image polygon.)*

In Investigation 1, you drew reflections over a line by folding paper or by using mirrors or transparent reflection tools. Why does the method you discovered in this problem give you the same results? *(When we folded paper or used a reflection tool to see the image of the original figure, every vertex was the same distance from the line of symmetry on the other side and in the same relative position. When we measure the distances, we make sure that each image point is the same distance from the line of reflection as the original point.)*

Have several students show their reflected figure from part B and the line of symmetry drawn for the figures in part C.

Explain how to draw an image of a figure using only a ruler and an angle ruler or protractor. *(Draw lines from each vertex on the polygon perpendicular to the line of reflection and extending beyond it. Measure the length of the line segment from each vertex to the line of reflection and measure an equal distance from the line of reflection to the opposite side. Mark the new points and connect them.)*

Part C asks you to reason in reverse: the reflection has been done, but the line of symmetry is not marked. Describe how to find the line of symmetry. *(Connect two corresponding vertices, and locate the midpoint of this line segment. The line of symmetry goes through this midpoint and is perpendicular to the line segment.)*

Part D asks students to complete a definition of a *line reflection*. Have two or three students share their answers. Each definition should capture the idea that point X' lies the same distance on the other side of the line of symmetry from point X on line segment XX', which is perpendicular to the line of symmetry.

When you are satisfied that students are comfortable with the basic concept of a line reflection, have them reconvene in their groups to complete the follow-up questions. For question 1, ask that each student make a drawing and then examine all the drawings and descriptions produced by their group.

Take time for a class discussion of the follow-up. Question 2 shows figures that intersect the line of reflection; questions 3 and 4 help to summarize and extend students' observations about line reflections.

When you perform a reflection over a line, could you match points other than the vertices of the figure with an image point? For example, what about a point inside the figure? Where is its image? *(on a line segment perpendicular to the line of reflection located the same distance from it and on the other side)*

What about a point outside the figure? *(It can be located in the same way.)* What about a point *on* the line of reflection? *(It is matched with itself.)*

So, can you tell me where any point on one side of a line of reflection would have an image? *(yes; always the same distance from and on the other side of the line of reflection)*

For the Teacher: Perpendicular Bisectors

Finding a perpendicular bisector is so useful in working with rigid motions of geometric figures that it is worth the time to introduce the term and help students develop some procedures for finding a perpendicular bisector. The most direct method is to locate the midpoint of the line segment and draw a line through it, perpendicular to the segment. However, you may want to discuss other methods as well.

Using a Straightedge and a Compass to Find a Perpendicular Bisector

To find the perpendicular bisector of the line segment connecting A to A', set the compass on point A to draw an arc that extends a bit beyond the midpoint of $\overline{AA'}$. Using that setting, draw an intersecting arc from point A'.

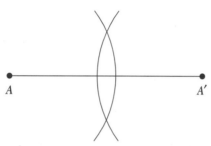

Draw the line connecting the points of intersection of the arcs. This line is the perpendicular bisector of $\overline{AA'}$. Note that the perpendicular bisector is the set of all points that are equidistant from A and A', which is why this procedure works.

Informal Methods for Finding the Perpendicular Bisector

Students can find the midpoint of $\overline{AA'}$ using several informal methods:

- folding the line segment in half

- marking the length of the segment on a strip of paper and folding the strip of paper in half

- using a transparent reflection tool

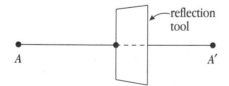

When A is matched with A', the reflection tool is on the midpoint.

After locating the midpoint, students can use a square corner, an angle ruler, or a protractor to draw a perpendicular line through the midpoint.

So, if you draw a figure and a line of reflection on a piece of paper, *every* point on the paper has an image on the other side of the line of reflection. You can think of the piece of paper as a plane that goes on forever. When you do a line reflection, *every* point in the plane has an image point. You can picture the plane flipping over the line of reflection, carrying the figure with it.

After discussing the follow-up, give students an opportunity to add to the definition of *line reflection* they wrote in part D.

2.2 • Describing Translations

In this problem and the next, students will explore many of the ideas considered in Problem 2.1, but involving different transformations. In this problem, they will investigate translations to observe properties that help to specify a translation in a way that will allow someone else to perform an identical translation.

Launch

Remind students of their work in Investigation 1, in which they explored translational symmetry.

> Look around the room. Do you see any patterns that show a translation?

Explain that the goals of this problem are similar to those of the last, but with a different kind of motion: a translation, or slide. Pose the open-ended question asked at the start of Problem 2.2:

> What is the relationship between a figure and its image under a translation?

Read through the problem with the class so that students know what is expected.

Pairs work well for this problem. Each student should do his or her own work, but check and discuss observations with a partner. Again, save the follow-up questions until after the summary discussion of the problem.

Explore

If students are having difficulty, encourage them to trace the polygon on tracing paper and then slide the tracing so that they can see the translation in progress. Sometimes observing this actual motion is all a student needs to make sense of translations.

For part B, students may need assistance to see that they can connect an original vertex with its image to determine the direction and distance, or magnitude, of the translation. Extending this line and marking off the same distance will locate the image of the first image vertex under the same translation performed a second time.

Summarize

Ask students to talk about their observations.

> What patterns did you see when you connected each vertex with its image? *(The lines were parallel and the same length.)*

> Does this make sense? *(Yes; under a translation, each point should slide the same distance and direction.)*

> How did you slide the image figure to perform the second translation?

Students will have various suggestions, but the idea that an original vertex, its image, and the image of that image vertex are on the same line should arise. The first image should be at the midpoint of the line connecting the original vertex and the second image.

Part C asks students to complete a definition of a *translation*. Have two or three students share their answers. Each definition should communicate the idea that the line segments connecting each point and image are all parallel and the same length. Each point has moved the same distance and in the same direction.

Have pairs explore the follow-up questions.

When students are ready, discuss the follow-up to ensure that everyone understands the ideas that an arrow can be drawn to indicate the length and direction of a translation and that it does not have to be in contact with the figure itself. In addition, the entire plane moves during a translation, so there are no fixed points.

2.3 • Describing Rotations

In this problem, students explore rotational transformation, again looking for properties that will help to specify a rotation so that someone else could perform an identical rotation. In this unit, rotations are always conducted in a counterclockwise direction.

Launch

Introduce the class to the topic of rotational transformation.

> Name some things that rotate.

Remind students of their work in Investigation 1, perhaps displaying some of the hubcaps and kaleidoscope designs from that investigation.

> In this problem, you will examine how to perform a rotation and how to find the image of an original figure under a rotation.

You might demonstrate a rotation at the overhead by drawing the same figure and marking the same point on two transparencies. Without explaining what you are doing, use the point as the center of rotation by anchoring it with a pin and rotating the top figure counterclockwise.

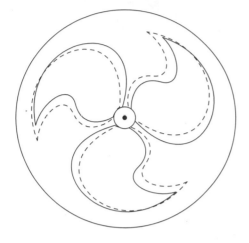

Anchor the transparencies through the center of rotation. Slowly turn the top transparency until it again coincides with the original figure.

Stop at some point in the rotation and ask:

How could we measure how far I have rotated the figure?

Let students offer their conjectures, and explain that they will learn about this idea as they work on the problem. Read through the problem with students, and have them explore it in pairs. Save the follow-up until after the summary.

Explore

If students are having trouble understanding the relationship between a figure and its image after a rotation, encourage the use of tracing paper or blank transparencies to help them understand the motion involved. Students can draw the original figure on tracing paper and mark the center of the rotation. By anchoring the tracing paper at the center of rotation with a pencil point and then rotating the paper, they can observe the "movement" of the image.

You might suggest that they imagine a vertex leaving a trail as it moves.

Suppose that as you rotate a figure, the vertices of the figure trace out their paths. What would the paths look like? *(a part of a circle, or an arc)*

Summarize

Call on students to share their observations from part A. They should express the idea that a vertex and its image are the same distance from the center of rotation and that all the angles formed by a vertex, the center of rotation, and the image of that vertex in a particular rotation have the same measure.

Have a student demonstrate how he or she performed the rotations in part B. If others have found another way to think about the rotations, give them a chance to share their methods.

Here are two common methods:

- Draw a line segment from point *A* to point *P*, and draw a 90° angle with this line segment as one side. Mark off the same distance along the other side, and mark the point *A'*. Do this for each point, and connect the resulting points.

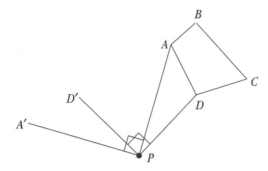

- Using point *Q* as the center, draw an arc from point *E*. Measure a 45° angle with an angle ruler or a protractor, using a line through points *E* and *Q* as one side of the angle. Mark the intersection of the angle with the arc as point *E'*. Do this for each point, and connect the resulting points.

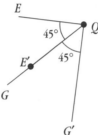

Have someone demonstrate how he or she rotated the image of the original polygon for one of the figures in part C. Students should have observed that each vertex "travels" along the same circle it traveled along during the first rotation, and that the angle of rotation has twice the measure of the original angle of rotation.

Part D asks students to complete a definition of a *rotation*. Have two or three students share their answers. Each definition should communicate the ideas that

- each vertex and its image are the same distance from the center of rotation

- the angles formed by each vertex, the center of rotation, and the image of the vertex are all the same within a given rotation

When students have a reasonable understanding of rotation, have them work with their partners on the follow-up. Note that Labsheet 2.3B contains two copies of the figures.

Question 2 is a challenge because students must reverse their thinking. By now, they know that each point and its image are the same distance from the center of rotation; from the problem, they should imagine a point rotating to its image position "riding" on the arc of a circle. From their work in Problem 2.1, students have encountered the concept of perpendicular bisectors. It will help to spend a few minutes exploring properties of points on a perpendicular bisector.

Draw a segment and its perpendicular bisector on the board or the overhead.

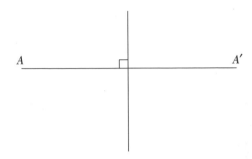

The line I have drawn goes through the midpoint of segment *AA′* and is perpendicular to it. What do we call such a line? *(a perpendicular bisector)*

If I mark a point *B* on the perpendicular bisector of segment *AA′*, what can you say for sure about point *B*?

If students are puzzled, mark point *B* and draw dashed lines from point *B* to points *A* and *A′*.

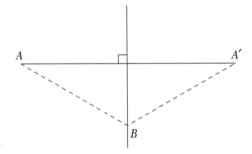

What can you say now? How do segments *AB* and *BA′* compare? *(They are the same length.)*

Would a different point on the perpendicular bisector show the same relationship?

Choose a point on the bisector at random and label it *C*. With the class, verify that this point shows the same relationship.

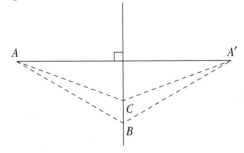

Do you think *any* point on the perpendicular bisector would show the same relationship? *(yes)* So, *every point on the perpendicular bisector is the same distance* from points *A* and *A′*.

Once students recognize that every point on the perpendicular bisector is the same distance from points *A* and *A′*, lead them in a discussion to discover this relationship: If a point is equidistant from points *A* and *A′*, that point must lie on the perpendicular bisector.

Begin a new drawing, starting with line segment *GG'*, as shown below. Using a compass, mark a point *K* that is the equidistant from points *G* and *G'*.

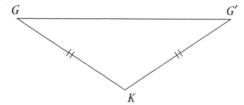

Next, draw a line segment from point *K* to the midpoint of segment *GG'* and label it point *H*.

Point *K* is the same distance from point *G* as it is from point *G'*. Line segment *KH* bisects segment *GG'*.

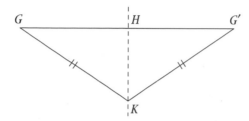

We know that lengths *KG* and *KG'* are equal and that lengths *GH* and *HG'* are equal. What can we conclude about triangles *GHK* and *G'HK*? *(They must be congruent, since they share side HK and thus have three pairs of equal sides.)*

Look at the angles *GHK* and *G'HK*. Since these add to a straight angle and are equal, each must have what measure? *(90°)*

So, we can conclude that segment *KH* is the perpendicular bisector of segment *GG'*.

Summarize what students now know.

We have learned that every point on the perpendicular bisector of segment *AA'* is equidistant from points *A* and *A'*. Also, we know that every point that is equidistant from points *A* and *A' must be* on the perpendicular bisector.

In question 2 of the follow-up, you are asked to find the center of rotation for a figure and its image. Think for a minute. If we look at a vertex, *A*, and its image, *A'*, under a rotation, what can we say about the distance from point *A* to the center of rotation and the distance from point *A'* to the center of rotation? *(They must be the same.)* So, what can we say about the *location* of the center of rotation? *(It must be on the perpendicular bisector of segment AA'.)*

See how you can use these ideas to finish the follow-up questions.

Be sure to discuss the follow-up to give students a chance to solidify their understanding of rotational transformations.

2.4 • Combining Transformations

In this problem, students analyze what happens when two reflections are performed in succession, the first reflection on the original figure and the second reflection on the image of that figure. Students consider two cases: one in which the two lines of reflection intersect, and the other in which the two lines of reflection are parallel. These ideas will be revisited in Investigation 4 when students explore combinations of symmetry transformations on geometric figures.

Launch

Review the concept of lines in a plane. To begin, ask that each student draw two lines on a sheet of paper. Then ask:

> How many of you drew lines that are parallel?
>
> Imagine that your lines extend forever in both directions. If your lines are *not* parallel, what must be true about them? *(The lines will eventually intersect.)*

Explain that students will be investigating what happens when they perform two transformations in succession, the first on the original figure and the second on the image of that figure.

> You will consider two cases: one in which the two lines of reflection *intersect,* and one in which the two lines of reflection are *parallel.* In each case, you will reflect a triangle over the first line and then reflect the triangle's image over the second line.
>
> Once you have performed the two transformations, you will analyze your drawing to see whether there is a *single* transformation you could perform to move the original triangle onto the final image.
>
> Does anyone have a conjecture at this stage about what you will discover?

Without commenting on students' ideas, you may want to list them for future reference. Have students perform the reflections on their own and then discuss with their partners what they think is happening. Save the follow-up questions until after the summary.

Explore

Caution students to work very carefully so that the second image they draw will be accurate.

Most students will simply perform the reflections by measuring, but some may want to try this procedure:

- Trace the original triangle and the first line of reflection, line 1, on tracing paper.

- Flip the tracing-paper drawing over, and match the line of reflection with the original line 1. Now you can see the first image.

- *Leaving the paper positioned on the first line of reflection*, draw the second line of reflection, line 2. This line will be on the opposite side of the paper from the drawing of the triangle.

- Flip the paper again. This time, match the second line of reflection with the original line 2. The tracing paper now shows the triangle in its second and final position.

- Press hard on the vertices of this triangle with the point of the pencil.

- Remove the tracing paper, and carefully draw the triangle. Label the vertices of this second image A'', B'', and C''.

Students can now place the tracing-paper image back over the original figure and determine what single transformation would "move" the original triangle to the final position.

If some students still need help trying to answer the question of what single transformation would produce the same result, suggest that they apply what they have learned about measuring distances and angles.

Summarize

Have students share what they have observed in the problem.

> What did you discover when you performed two reflections over the pair of *intersecting* lines in part A? *(When we measured the distance of the original vertex and its image from the point of intersection, it was the same in every case. When we measured the angles formed by each original vertex, the intersection point, and the final image of the vertex, the angle measures were all the same.)*
>
> What kind of transformation does this resemble? *(a rotation)*
>
> Tell me more about this rotation.

The center of rotation is the point of intersection of the lines. The amount of rotation is twice the measure of the angle between the intersecting lines of reflection. If students have not discovered this second relationship, suggest that they compare the measure of the angle formed by line 1 and line 2 to the angle of rotation.

Now review part B.

> What did you discover when you performed two reflections over the pair of *parallel* lines in part B? *(When we measured the distances from each vertex to its final image, they were all the same.)*
>
> What kind of transformation does this resemble? *(a translation)*
>
> Tell me more about this translation. Exactly what translation will produce the same final image?

If students do not describe the single translation accurately, model what they have said to encourage them to be more precise. For example, students might say, "You must slide the figure toward the lines of reflection." If so, demonstrate a transformation that is not perpendicular to the lines of reflection, and ask whether this is what they mean.

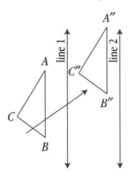

Can you be more specific? *(The slide has to be in a direction perpendicular to the lines of reflection.)*

And how far must I translate the figure?

Help students confirm that the translation is equal to twice the distance between the lines of reflection.

Review what students have learned about these two cases:

■ A reflection of a figure over two intersecting lines is equivalent to a rotation with the point of intersection as the center and a length, or magnitude, that is twice the measure of the angle between the lines.

■ A reflection of a figure over two parallel lines is equivalent to a translation in a direction perpendicular to the parallel lines and for a magnitude of twice the distance between them.

Put students back into their pairs to work on the follow-up questions. You may want to assign the questions as homework.

The follow-up questions reverse what you know and what you are to do. In 2 and 3, you are given the original figure and the second image after two reflections. You have to find two lines that can be used as lines of reflection to "move" the original figure onto the final image. Parts 4 and 5 ask about combining two translations or two rotations.

Take time to discuss the follow-up when students have had an opportunity to work on it. If no one comes up with a strategy for solving question 2, ask the following questions and let students try again.

How can you tell how far apart the parallel lines must be? *(Measure the length of segment AA", or BB", or CC"; the parallel lines must be half this distance apart.)*

What relationship must the parallel lines have to the segments *AA"*, *BB"*, and *CC"*? *(They must be perpendicular to these segments.)*

For question 3, students need to recognize that reflecting over two intersecting lines is equivalent to a rotation.

> What single transformation will move polygon *ABCD* to *A"B"C"D"*? *(A rotation?)* How can you find the center of that rotation? *(Use perpendicular bisectors.)*

> So, what relationships must exist between the two lines we are looking for? *(They must intersect at the center of rotation, and the measure of the angle between them must be half that of the angle of rotation.)*

Have students share some of their drawings for questions 2 and 3 so they can see the variety of ways to solve these problems. Here are three ways to draw two parallel lines for the figures in question 2:

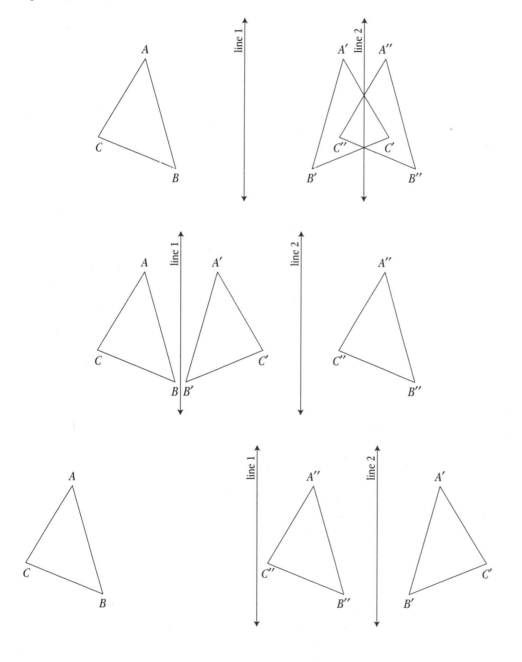

Here are three ways to draw two intersecting lines for the figures in question 3:

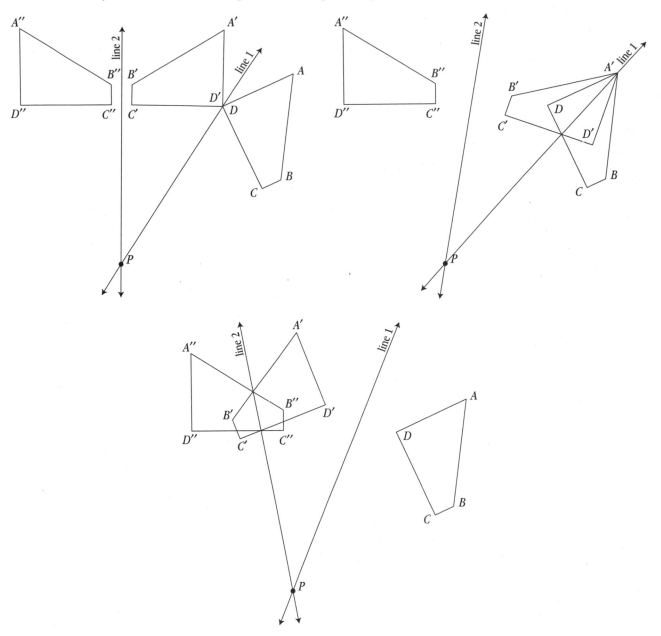

Additional Answers

Answers to Problem 2.1

C. Draw line segments joining each pair of corresponding vertices (a vertex of one figure and its image vertex in the other). Measure to find the midpoint of each line segment. The line that joins the midpoints is the line of symmetry. (Note: Students might also use an angle ruler or a protractor to draw the line of symmetry perpendicular to and through the midpoint of a line segment joining a pair of corresponding vertices.)

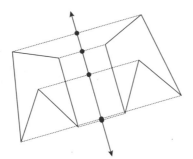

Answers to Problem 2.1 Follow-Up

2. a.

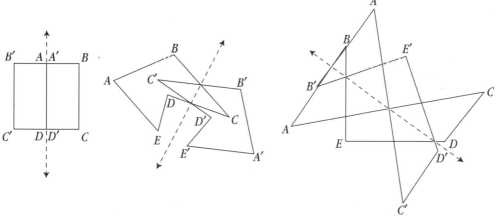

b. The final drawing in each case has reflectional symmetry. The part of the final drawing on one side of the line of reflection corresponds exactly to the part of the drawing on the other side.

Answers to Problem 2.2

A. 1.

diagram 1   diagram 2

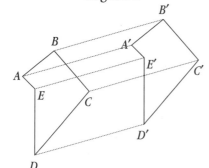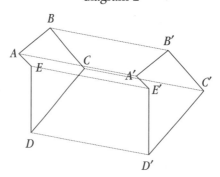

2. All the connecting line segments are parallel and of the same length.

B. 1.

diagram 1

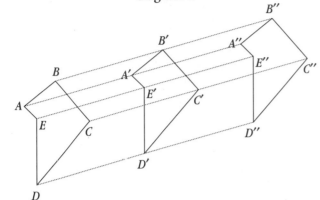

diagram 2

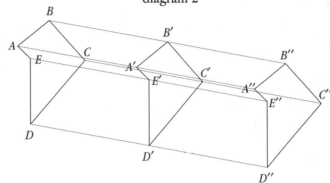

2. Polygon *A"B"C"D"E"* is twice as far from the original, in the same direction, as polygon *A'B'C'D'E' is*. The first image's vertices are the midpoints of the line segments connecting an original vertex and its second image.

3. The final drawing has the beginning of translational symmetry. (**Teaching Tip:** Students will likely say the drawing has translational symmetry. However, because the design does not repeat forever, it technically does not have translational symmetry. You may want to further explore this idea with your class.)

Answers to Problem 2.3

B. 1.

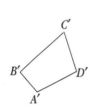 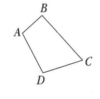 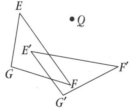

2. In the quadrilateral, each vertex travels along a circle centered at point *P* and moves through an angle of 90°. In the triangle, each vertex travels along a circle centered at point *Q* and travels through an angle of 45°.

C. Each vertex of the final image is the same distance from the center as the original vertex is, and the angle formed has twice the measure of the original angle of rotation.

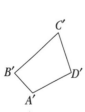 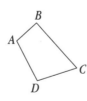

 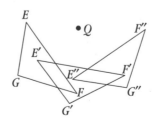

D. Possible answer: A *rotation* of *d* degrees about a point *P* matches any point *X* on a figure to an image point *X'* so that the measure of angle *XPX'* is *d* degrees and the distance *XP* equals the distance *X'P.*

Answers to Problem 2.3 Follow-Up

2. a. The center of rotation is marked as point *P* below. Explanations will vary. Students might draw the perpendicular bisectors of the lines connecting two pairs of vertices and images. The intersection point of these perpendicular bisectors is the center of rotation. Some students may use trial and error. As long as they can explain how to test whether the point is the center of rotation, this is all right at this stage.

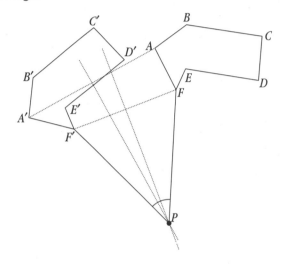

  b. The angle of rotation is 47°, which can be found by drawing and measuring the angle formed by a vertex, the center of rotation, and the image of that vertex.

3. Each of these points is on the perpendicular bisector of the segment joining a vertex and its image.

4. The distances are the same.

5. No; none of these drawings have rotational symmetry. **Teaching Tip:** The rotation of a given figure will produce a symmetric original-image design only if the measure of the angle of rotation is a factor, *k*, of 360 and the design consists of 360 ÷ *k* figures. You may want to explore this idea with the class. For example, the quadrilateral in part B can be rotated once more to produce a drawing with rotational symmetry, and the triangle can be rotated five more times to produce a drawing with rotational symmetry. The quadrilateral is rotated 90° and the final drawing thus contains 360 ÷ 90 = 4 figures. The triangle is rotated 45° and the final drawing thus contains 360 ÷ 45 = 8 figures.

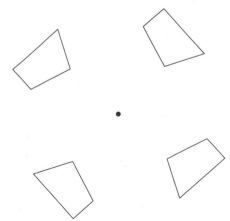

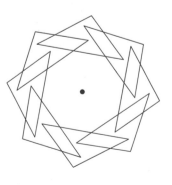

6. Possible answer: Tools for measuring angles and lengths are useful for finding the location of each vertex of the image. A compass can be useful for marking image points. A straightedge is useful for connecting the points of the image.

7. The center of rotation is the only fixed point.

Answers to Problem 2.4

B. 1. Possible answer: The result might be equivalent to a single translation.

2. Answers will vary. The measures of the line segments between any point and its final image are equal. Because these line segments are perpendicular to the lines of reflection, this transformation can be done as a single translation. The slide must be in a direction perpendicular to the lines of reflection and for a distance of twice the distance between the lines.

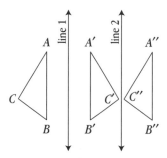

Answers to Problem 2.4 Follow-Up

2. Drawings will vary. Possible explanations: Draw a line from point A to point A″, measure its length, and then draw two parallel lines that are perpendicular to segment AA″ and separated by half the length of segment AA″. Or, draw a line from A to point A″. Then draw the first line of symmetry somewhere between the triangles and perpendicular to segment AA″. Draw the reflection of triangle ABC over that line. Finally, draw the line of symmetry between triangle A′B′C′ and triangle A″B″C″.

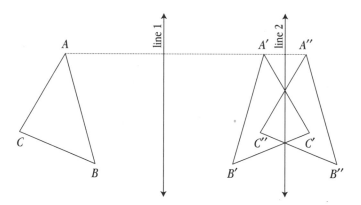

3. Possible explanations: Find the angle of rotation and the center of rotation, and then take half that angle to draw two lines of reflection. *Or,* draw the perpendicular bisectors of segments *CC"* and *DD",* which intersect in the center of rotation. Draw any two lines through this point that form an angle whose measure is equal to half that of the angle formed by vertex *A,* the point of intersection, and vertex *A".*

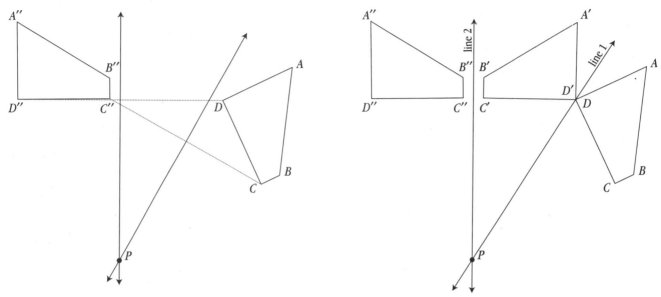

Find the center of rotation, P.   *Draw two lines of reflection.*

4. Two translations are equivalent to a single translation. The single translation can be found by drawing an arrow to represent the first translation and drawing a second arrow from the end of the first arrow to represent the second translation. A single arrow drawn from the start of the first arrow to the end of the second arrow indicates the equivalent, single translation.

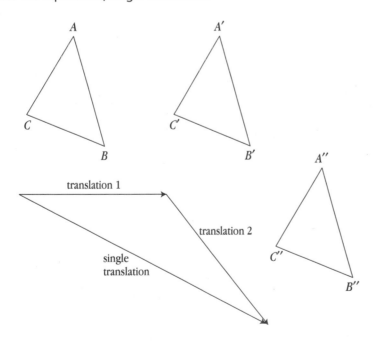

ACE Answers

Applications

11.

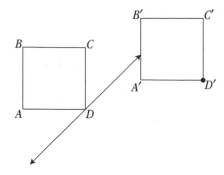

Connections

17. Possible answer: The design below has translational symmetry as indicated by the arrow and reflectional symmetry over the vertical line.

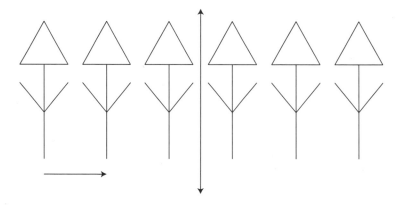

20. Possible answer: The design below has translational symmetry as indicated by the arrow, but it has no reflectional or rotational symmetry because there is no way to fold the design so that the halves match and there is no way to rotate it so that it looks like the original design.

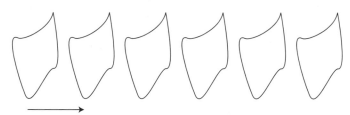

Transforming Coordinates

This investigation connects the mathematical language for describing symmetry transformations of geometric figures to their coordinate representations. Building on their prior experiences, students are asked to formulate clear algebraic descriptions of symmetry transformations in terms of matching the coordinates of points in original figures to those of their symmetric images.

In the grade 7 unit *Stretching and Shrinking*, students represented transformations by writing rules for the coordinates of a point under a transformation. In that unit, they wrote rules for producing *similar* figures; in this investigation, they write rules for producing *congruent* figures.

In Problem 3.1, Writing Rules for Reflections; Problem 3.2, Writing Rules for Translations; and Problem 3.3, Writing Rules for Rotations, students write rules for transforming a point (x, y) to its image point. The explorations are set in the context of writing computer instructions for drawing line segments, which focuses attention on what happens to the coordinates of a point under the desired transformation. In Problem 3.4, Relating Symmetry to Congruence, students apply their knowledge of symmetry transformations to determine whether two figures are congruent. This will help them to connect the ideas of symmetry, transformations, and the familiar geometric concept of congruence.

Mathematical and Problem-Solving Goals

- **To write directions for drawing figures composed of line segments**

- **To analyze the vertices of a figure under a transformation and to specify translations with coordinate rules**

- **To recognize that a transformation of the form $(x, y) \rightarrow (x + a, y + b)$ is a translation of point (x, y) a units in the x direction and b units in the y direction**

- **To specify rotations of 90°, 180°, 270°, and 360° with coordinate rules**

- **To specify reflections over the x-axis, the y-axis, and the line y = x**

- **To combine transformations to find single, equivalent transformations**

- **To understand the relationship between symmetry transformations and congruence**

Materials

Problem	For students	For the teacher
All	Graphing calculators, centimeter or quarter-inch grid paper	Transparencies: 3.1A to 3.4 (optional), overhead graphing calculator (optional)
3.1	Labsheet 3.1A (optional; 1 per group), Labsheet 3.1B (1 per group), tracing paper	Computer graphics software (optional), transparent centimeter grid (optional)
3.3	Labsheet 3.3 (optional; 1 per group), tracing paper	Transparent centimeter grids (optional)
3.4	Labsheet 3.4 (1 per student), rulers (1 per student), angle rulers or protractors (1 per student)	Transparent ruler (optional)
ACE	Labsheets 3.ACE1, 3.ACE2, and 3.ACE3 (1 ea. per student)	

3.1

Writing Rules for Reflections

Launch

- Talk about drawing designs by specifying endpoints of line segments.

- Have groups of two to four explore the problem and follow-up.

Explore

- Students may need help drawing the line $y = x$ in the problem and writing the first rule in the follow-up.

Summarize

- Have students discuss the patterns they saw and the general rules they wrote.

- Assess understanding by having students write a rule for reflection over a different line.

Transforming Coordinates

In this investigation, you will explore transformations of figures drawn on coordinate grids. By looking for patterns in your results, you will be able to write some general algebraic rules for transforming a point (x, y) under reflections, translations, and rotations.

3.1 **Writing Rules for Reflections**

The drawing screen in many computer geometry programs is considered to be a coordinate grid. You can create designs by specifying the endpoints of line segments.

The flag below consists of three segments. The commands for creating the flag in a particular geometry program are shown next to the screen. The commands tell the computer to draw segments between the specified endpoints.

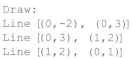

```
Draw:
Line [(0,-2), (0,3)]
Line [(0,3), (1,2)]
Line [(1,2), (0,1)]
```

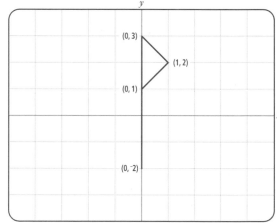

Assignment Choices

ACE questions 1–3, 9–11, 15, and unassigned choices from earlier problems

Think about this!

Is there a different set of commands that would create the same flag?

What commands would create a square centered at the origin?

What commands would create a nonsquare rectangle?

Problem 3.1

A. Suppose you want to re-create the flag below using the geometry program that drew the flag on page 42. Copy and complete the commands to create a set of instructions for drawing the flag.

```
Draw:
Line [( , ), ( , )]
Line [( , ), ( , )]
Line [( , ), ( , )]
```

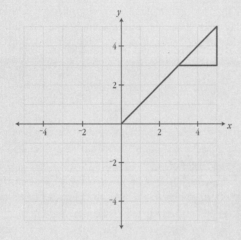

B. Write a set of commands that would draw the image of this flag under a reflection over the y-axis.

C. Write a set of commands that would draw the image of this flag under a reflection over the x-axis.

D. Write a set of commands that would draw the image of this flag under a reflection over the line y = x.

Answers to Problem 3.1

A possible set of commands is given for each part.

```
A.  Draw:
    Line [(0,0), (5,5)]
    Line [(3,3), (5,3)]
    Line [(5,3), (5,5)]
```

```
B.  Draw:
    Line [(0,0), (-5,5)]
    Line [(-3,3), (-5,3)]
    Line [(-5,3), (-5,5)]
```

```
C.  Draw:
    Line [(0,0), (5,-5)]
    Line [(3,-3), (5,-3)]
    Line [(5,-3), (5,-5)]
```

```
D.  Draw:
    Line [(0,0), (5,5)]
    Line [(3,3), (3,5)]
    Line [(3,5), (5,5)]
```

■ Problem 3.1 Follow-Up

The diagram below appears on Labsheet 3.1B. Use this diagram to complete parts 1–4.

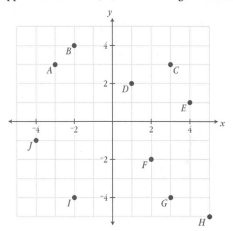

1. List the coordinates of the labeled points.

2. **a.** Indicate the images of points *A–J* under a reflection over the *y*-axis. Use the label *A′* for the image of point *A*, *B′* for the image of *B*, and so on.

   b. List the coordinates of points *A′–J′*.

   c. Compare the coordinates of each original point with the coordinates of its image. Use the pattern you see to complete this general rule for finding the image of any point (x, y) under a reflection over the *y*-axis:

   $$(x, y) \rightarrow (\quad , \quad)$$

3. **a.** Indicate the images of points *A–J* under a reflection over the *x*-axis. Use the label *A″* for the image of point *A*, *B″* for the image of *B*, and so on.

   b. List the coordinates of points *A″–J″*.

   c. Compare the coordinates of each original point with the coordinates of its image. Use the pattern you see to complete this general rule for finding the image of any point (x, y) under a reflection over the *x*-axis:

   $$(x, y) \rightarrow (\quad , \quad)$$

4. **a.** Indicate the images of points *A–J* under a reflection over the line $y = x$. Use the label *A‴* for the image of point *A*, *B‴* for the image of *B*, and so on.

   b. List the coordinates of points *A‴–J‴*.

   c. Compare the coordinates of each original point with the coordinates of its image. Use the pattern you see to complete this general rule for finding the image of any point (x, y) under a reflection over the line $y = x$:

   $$(x, y) \rightarrow (\quad , \quad)$$

Answers to Problem 3.1 Follow-Up

1. See page 58k.

2a, b. See page 58k.

2c. $(x, y) \rightarrow (^{-}x, y)$

3a, b. See page 58k.

3c. $(x, y) \rightarrow (x, ^{-}y)$

4a, b. See page 58l.

4c. $(x, y) \rightarrow (y, x)$

3.2 Writing Rules for Translations

The designs below have translational symmetry. In this problem, you will explore the computer commands needed to create these designs.

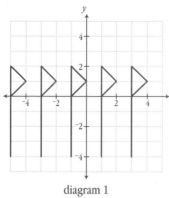

diagram 1

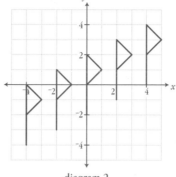

diagram 2

Problem 3.2

A. 1. In diagram 1, the left-most flag can be drawn with these commands:

```
Draw:
Line [(-5,-4), (-5,2)]
Line [(-5,2), (-4,1)]
Line [(-4,1), (-5,0)]
```

These commands draw the vertical segment, then the upper slanted segment, and finally the lower slanted segment. Write sets of commands for drawing the other four flags in diagram 1. Each set of commands should draw the segments in the same order as the commands for the original flag.

2. Compare the commands for the five flags. Describe a pattern that relates the coordinates of each flag to the coordinates of the flag to its *right*.

3. Describe a pattern that relates the coordinates of each flag to the coordinates of the flag to its *left*.

B. 1. Write a set of commands for drawing the left-most flag in diagram 2. Then write comparable instructions for drawing the other four flags.

2. Compare the commands for the five flags. Describe a pattern that relates the coordinates of each flag to the coordinates of the flag to its *right*.

3. Describe a pattern that relates the coordinates of each flag to the coordinates of the flag to its *left*.

Investigation 3: Transforming Coordinates **45**

At a Glance

Grouping:
pairs

Launch

- Talk about the two designs and the symmetry they show.

- Have pairs explore the problem and follow-up.

Explore

- Focus students on relating the coordinates between the original and the first image.

- Point out that the follow-up does not specify the *length* of the translation.

Summarize

- Introduce the idea of using a variable in coordinate rules for translations.

- As a class, explore the relationship between slopes and rules for translations and what is true for all rules that specify translations.

Answers to Problem 3.2

A. 1. See page 58l.

2. To find the coordinates of the flag to the right of a given flag, add 2 to each *x*-coordinate and keep the *y*-coordinates the same.

3. To find the coordinates of the flag to the left of a given flag, subtract 2 from each *x*-coordinate and keep the *y*-coordinates the same.

B. 1. See page 58l.

2. To find the coordinates of the flag to the right of a given flag, add 2 to each *x*-coordinate and 1 to each *y*-coordinate.

3. To find the coordinates of the flag to the left of a given flag, subtract 2 from each *x*-coordinate and 1 from each *y*-coordinate.

Assignment Choices

ACE questions 6–8, 17, 19–21, and unassigned choices from earlier problems

■ **Problem 3.2 Follow-Up**

In 1–3, refer to the figure below.

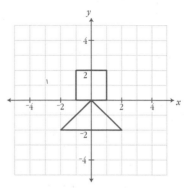

1. Suppose you want to create a design with translational symmetry by sliding this basic figure along the *x*-axis. Consider different designs you could make by using translations of various lengths.
   a. In any design created by translating the figure along the *x*-axis, how are the coordinates of the original figure related to the coordinates of the figure to its right?
   b. How are the coordinates of the second copy to the right of the original related to the coordinates of the first copy to the right? How are the coordinates of the second copy related to the coordinates of the original?
   c. How are the coordinates of copies to the left of the original related to the coordinates of the original?

2. Consider the designs you could make by sliding this basic figure along the *y*-axis.
   a. In any such design, how are the coordinates of the original figure related to the coordinates of the figure directly above it?
   b. How are the coordinates of the second copy above the original related to the coordinates of the copy directly below it? How are the coordinates of the second copy related to the coordinates of the original?
   c. How are the coordinates of copies below the original related to the coordinates of the original?

3. Now consider designs created by sliding the figure along the slanted, or *oblique*, lines $y = x$, $y = {}^-x$, and $y = \frac{1}{2}x$. Make sketches of such designs, and examine how the coordinates change from one figure to the next. For each line, tell what happens to a point on the figure as the figure is translated along the line.

Answers to Problem 3.2 Follow-Up

1. a. The *x*-coordinate of the figure to the right of a given figure is *x* plus the length of the translation; the *y*-coordinate does not change.

   b. The *x*-coordinate of the figure second to the right of a given figure is the *x*-coordinate of the first copy to the right plus the length of the translation. It is also the *x*-coordinate of the original figure plus twice the amount of the translation. In both cases, the *y*-coordinate does not change.

   c. The *x*-coordinate of any figure to the left of a given figure is *x* minus the length of the translation; the *y*-coordinate does not change.

2–5. See page 58m.

4. Look back at the translations you explored in questions 1–3. In each case, think about the rules for finding the image of any point (x, y). How are the rules similar?

5. Which of these rules describe translations? Explain.
  a. $(x, y) \rightarrow (x + 2, y - 3)$
  b. $(x, y) \rightarrow (2x, y)$
  c. $(x, y) \rightarrow (x + 1, 3y)$
  d. $(x, y) \rightarrow (x - 2, y + 1)$

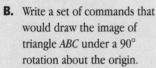

3.3 Writing Rules for Rotations

You have explored the rules for transforming a point (x, y) to its image under reflections and translations. Writing rules for rotations is more difficult. In this problem, you will explore a few simple cases.

Problem 3.3

A. Copy and complete the commands below to create a set of instructions for drawing triangle *ABC*.

```
Draw:
Line [( , ), ( , )]
Line [( , ), ( , )]
Line [( , ), ( , )]
```

B. Write a set of commands that would draw the image of triangle *ABC* under a 90° rotation about the origin.

C. Write a set of commands that would draw the image of triangle *ABC* under a 180° rotation about the origin.

D. Write a set of commands that would draw the image of triangle *ABC* under a 270° rotation about the origin.

E. Write a set of commands that would draw the image of triangle *ABC* under a 360° rotation about the origin.

3.3

Writing Rules for Rotations

At a Glance

Grouping:
pairs

Launch

■ Introduce the goal of writing commands to specify the image of a figure under a rotation.

■ Have pairs explore the problem and follow-up.

Explore

■ If students are struggling, suggest that they sketch each rotation image.

■ Help students think about follow-up 3 and 4 in terms of coordinates.

Summarize

■ Help the class understand what is happening to the coordinates by studying the rotation of a grid.

Answers to Problem 3.3

A possible set of commands is given for each part.

A. Draw:
```
Line [(1,4), (4,2)]
Line [(4,2), (2,0)]
Line [(2,0), (1,4)]
```

B. Draw:
```
Line [(-4,1), (-2,4)]
Line [(-2,4), (0,2)]
Line [(0,2), (-4,1)]
```

C. Draw:
```
Line [(-1,-4), (-4,-2)]
Line [(-4,-2), (-2,0)]
Line [(-2,0), (-1,-4)]
```

D. Draw:
```
Line [(4,-1), (2,-4)]
Line [(2,-4), (0,-2)]
Line [(0,-2), (4,-1)]
```

E. Draw:
```
Line [(1,4), (4,2)]
Line [(4,2), (2,0)]
Line [(2,0), (1,4)]
```

Assignment Choices

ACE questions 4, 12–14, 18, 22–24, and unassigned choices from earlier problems

Relating Symmetry to Congruence

Grouping:
individuals

Launch

- Review what students know about congruent figures.

- Read the problem and follow-up as a class.

- Have students explore the problem and follow-up.

Explore

- As an added activity, have students draw congruent figures and have a partner find translations to get from one to the other.

Summarize

- Discuss the problem, having students demonstrate transformations that would match the pentagons.

■ **Problem 3.3 Follow-Up**

1. a. Organize your results from Problem 3.3 as shown below to indicate the images of the vertices under each rotation.

90° rotation	180° rotation	270° rotation	360° rotation
$A(1, 4) \rightarrow (\ ,\ )$	$A(1, 4) \rightarrow (\ ,\ )$	$A(1, 4) \rightarrow (\ ,\ )$	$A(1, 4) \rightarrow (\ ,\ )$
$B(4, 2) \rightarrow (\ ,\ )$	$B(4, 2) \rightarrow (\ ,\ )$	$B(4, 2) \rightarrow (\ ,\ )$	$B(4, 2) \rightarrow (\ ,\ )$
$C(2, 0) \rightarrow (\ ,\ )$	$C(2, 0) \rightarrow (\ ,\ )$	$C(2, 0) \rightarrow (\ ,\ )$	$C(2, 0) \rightarrow (\ ,\ )$

b. Describe what happens to the vertices of triangle *ABC* under each rotation. Be sure to discuss how each rotation affects the coordinates of the points.

2. Complete the following statements:
  a. A rotation of 90° moves point (x, y) to point $(\ ,\ )$.
  b. A rotation of 180° moves point (x, y) to point $(\ ,\ )$.
  c. A rotation of 270° moves point (x, y) to point $(\ ,\ )$.
  d. A rotation of 360° moves point (x, y) to point $(\ ,\ )$.

3. What single transformation produces the same final image of triangle *ABC* as a reflection over the *y*-axis followed by a reflection over the *x*-axis? Explain.

4. What single transformation produces the same final image of triangle *ABC* as a reflection over the *y*-axis followed by a reflection over the line $y = x$? Explain.

3.4 **Relating Symmetry to Congruence**

Congruent figures have the same size and shape. Congruent figures can also be defined using the language of symmetry transformations: *Two figures are congruent if one is an image of the other under a reflection, a translation, a rotation, or some combination of these transformations.* Put more simply, two figures are congruent if you can flip, slide, or turn one figure so that it fits exactly on the other.

Pentagons *ABCDE* and *PQRST* are congruent. You will look at these pentagons more carefully in the problem.

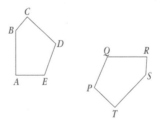

Assignment Choices

ACE questions 5, 16, 25–27, and unassigned choices from earlier problems

Assessment

It is appropriate to use Check-Up 2 after this problem.

Answers to Problem 3.3 Follow-Up

1. a.

90° rotation	180° rotation	270° rotation	360° rotation
$A (1, 4) \rightarrow (^-4, 1)$	$A (1, 4) \rightarrow (^-1, ^-4)$	$A (1, 4) \rightarrow (4, ^-1)$	$A (1, 4) \rightarrow (1, 4)$
$B (4, 2) \rightarrow (^-2, 4)$	$B (4, 2) \rightarrow (^-4, ^-2)$	$B (4, 2) \rightarrow (2, ^-4)$	$B (4, 2) \rightarrow (4, 2)$
$C (2, 0) \rightarrow (0, 2)$	$C (2, 0) \rightarrow (^-2, 0)$	$C (2, 0) \rightarrow (0, ^-2)$	$C (2, 0) \rightarrow (2, 0)$

b. *Under a 90° rotation,* the *x*- and *y*-coordinates exchange places, and then the sign of the new *x*-coordinate is reversed. *Under a 180° rotation,* the signs of the *x*- and *y*-coordinates are reversed. *Under a 270° rotation,* the *x*- and *y*-coordinates exchange places, and then the sign of the new *y*-coordinate is reversed. *Under a 360° rotation,* the *x*- and *y*-coordinates are unchanged.

2. a. $(^-y, x)$ **b.** $(^-x, ^-y)$ **c.** $(y, ^-x)$ **d.** (x, y)

3, 4. See page 58m.

Problem 3.4

Pentagons *ABCDE* and *PQRST* are reproduced on Labsheet 3.4.

A. If you made a copy of one of the pentagons and fit it exactly on the other, which vertices would match?

B. Which pairs of sides in pentagons *ABCDE* and *PQRST* are the same length?

C. Which pairs of angles in pentagons *ABCDE* and *PQRST* are the same size?

D. What combination of reflections, rotations, and translations would move pentagon *ABCDE* to fit exactly on pentagon *PQRST*? Is there more than one possible combination? Make sketches to show your ideas.

■ Problem 3.4 Follow-Up

1. Drawing a diagonal of a rectangle creates two congruent triangles.

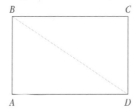

a. If you make a copy of triangle *ABD* and place it exactly on the other triangle, which vertices will match?

b. What single reflection, rotation, or translation would match one of the triangles exactly with the other?

c. Which pairs of sides and angles in the two triangles are congruent?

2. a. How could you determine whether these circles are congruent without making copies of them or performing transformations?

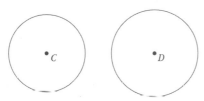

b. What transformations could you perform to move one of the circles onto the other to check for congruence?

Answers to Problem 3.4

A. *A* and *T, B* and *S, C* and *R, D* and *Q, E* and *P*

B. *AB* and *TS, BC* and *SR, CD* and *RQ, DE* and *QP, EA* and *PT*

C. ∠*A* and ∠*T*, ∠*B* and ∠*S*, ∠*C* and ∠*R*, ∠*D* and ∠*Q*, ∠*E* and ∠*P*

D. Pentagon *ABCDE* could be translated toward pentagon *PQRST* until points *A* and *T* coincided. Pentagon *ABCDE* could then be reflected over a line as shown here. Other combinations would work; for example, pentagon *ABCDE* could be reflected and then translated.

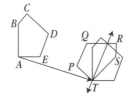

Answers to Problem 3.4 Follow-Up

See page 58n.

Answers

Applications

1a. Possible answer:
Draw:
Line [(1,-3), (1,3)]
Line [(1,3), (4,3)]
Line [(4,3), (4,0)]
Line [(4,0), (2,0)]
Line [(2,0), (4,-3)]

1b. See drawing below right. Possible answer:
Draw:
Line [(-1,-3), (-1,3)]
Line [(-1,3), (-4,3)]
Line [(-4,3), (-4,0)]
Line [(-4,0), (-2,0)]
Line [(-2,0), (-4,-3)]

2a. Possible answer:
Draw:
Line [(-4,-4), (0,4)]
Line [(0,4), (4,-4)]
Line [(-2,0), (2,0)]

2b. See drawing below right. Possible answer:
Draw:
Line [(-4,4), (0,-4)]
Line [(0,-4), (4,4)]
Line [(-2,0), (2,0)]

As you work on these ACE questions, use your calculator whenever you need it.

Applications

In 1–3, refer to the figures below. These figures are reproduced on Labsheet 3.ACE1.

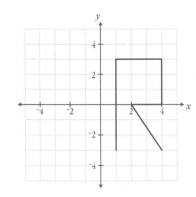
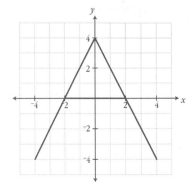

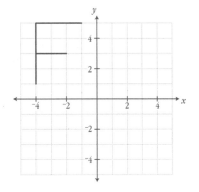

1b.

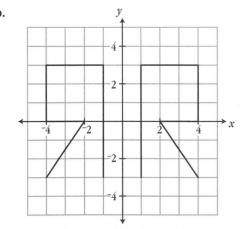

2b.

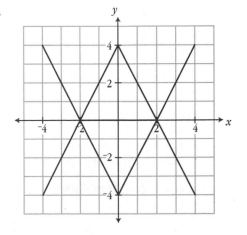

1. a. Write a set of computer commands for drawing the letter R shown.

   b. Write a set of commands for drawing the image of this R under a reflection over the *y*-axis. Use your instructions to draw the image.

2. a. Write a set of computer commands for drawing the letter A shown.

   b. Write a set of commands for drawing the image of this A under a reflection over the *x*-axis. Use your instructions to draw the image.

3. a. Write a set of computer commands for drawing the letter F shown.

   b. Write a set of commands for drawing the image of this F under a reflection over the line *y = x*. Use your instructions to draw the image.

4. Draw a figure on a coordinate grid. Perform one transformation on your original figure and a second transformation on its image. Is there a single transformation that will produce the same final result? Explain.

5. Diagonal *QS* divides the parallelogram into two congruent triangles.

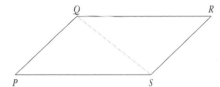

   a. If you make a copy of triangle *PQS* and place it exactly on the other triangle, which vertices will match?

   b. What single reflection, translation, or rotation would match one of the triangles exactly with the other?

   c. Which pairs of sides and angles in the two triangles are congruent?

3b.

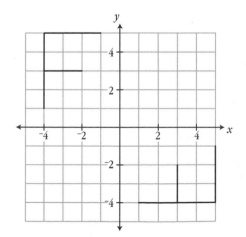

3a. Possible answer:
```
Draw:
[(-1,5), (-4,5)]
[(-4,5), (-4,1)]
[(-4,3), (-2,3)]
```

3b. See drawing below left. Possible answer:
```
Draw:
[(5,-1), (5,-4)]
[(5,-4), (1,-4)]
[(3,-4), (3,-2)]
```

4. Answers will vary. If students perform reflections over intersecting lines or rotations about the same center, they can find a single, equivalent rotation. If they perform reflections over parallel lines, they can find a single, equivalent translation. If they combine two translations, they can find a single, equivalent translation. If they combine a reflection and a translation, they can find a single, equivalent glide reflection. There are other combinations that students might try.

5a. *P* and *R*, *Q* and *S*, *S* and *Q*

5b. a 180° rotation about the midpoint of segment *QS*

5c. sides: *PQ* and *RS*, *PS* and *RQ*, *QS* and *SQ*; angles: ∠*P* and ∠*R*, ∠*PQS* and ∠*RSQ*, ∠*PSQ* and ∠*RQS*

6. (⁻2, 5)

7. (0, 1)

8. (0, ⁻2)

9. (⁻1, 2)

10. (4, ⁻1)

11. (⁻2, 2)

12. (4, 3)

In 6–15, refer to the figure below. This figure is reproduced on Labsheet 3.ACE1.

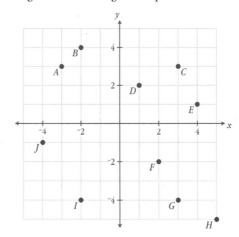

6. What are the coordinates of the image of point A under a translation that moves point (1, 2) onto point (2, 4)?

7. What are the coordinates of the image of point B under a translation that moves point (1, 2) onto point (3, ⁻1)?

8. What are the coordinates of the image of point C under a translation that moves point (1, 2) onto point (⁻2, ⁻3)?

9. What are the coordinates of the image of point D under a reflection over the y-axis?

10. What are the coordinates of the image of point E under a reflection over the x-axis?

11. What are the coordinates of the image of point F under a reflection over the line y = x?

12. What are the coordinates of the image of point G under a 90° rotation about the origin?

13. What are the coordinates of the image of point *H* under a 180° rotation about the origin?

14. What are the coordinates of the image of point *I* under a 270° rotation about the origin?

15. What are the coordinates of the final image of point *J* under a reflection over the *x*-axis followed by a reflection over the *y*-axis?

16. The figure below appears on Labsheet 3.ACE2.

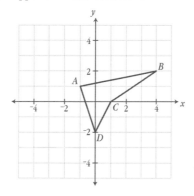

a. Draw the final image that results from rotating polygon *ABCD* 90° about the origin and then reflecting the image over the *x*-axis.

b. Draw the final image that results from reflecting polygon *ABCD* over the *x*-axis and then rotating the image 90° about the origin.

c. Are the final images you found in parts a and b the same? Why or why not?

17. What single transformation is equivalent to a 90° rotation about the origin followed by a 270° rotation about the origin?

18. What single transformation is equivalent to a reflection over the *y*-axis, followed by a reflection over the *x*-axis, followed by another reflection over the *y*-axis?

---

13. (⁻5, 5)

14. (⁻4, 2)

15. (4, 1)

16a. The image, polygon *A'B'C'D'*, is shown below left.

16b. The image, polygon *A'B'C'D'*, is shown below left.

16c. A rotation of 90° followed by a reflection over the *x*-axis takes point (*x, y*) to point (⁻*y, x*) and then takes (⁻*y, x*) to point (⁻*y,* ⁻*x*). A reflection over the *x*-axis followed by a rotation of 90° takes point (*x, y*) to point (*x,* ⁻*y*) and then takes point (*x,* ⁻*y*) to point (*y, x*). They are not the same, so order matters.

17. a 360° rotation about the origin

18. a reflection over the *x*-axis

16a.

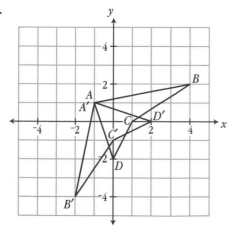

16b.

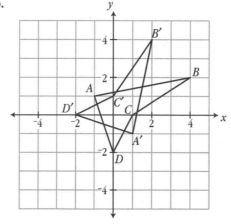

Connections

19. reflectional symmetry over a vertical line through the center; reflectional symmetry over a horizontal line through the center; rotational symmetry with a 180° angle of rotation

20. rotational symmetry with a 60° angle of rotation

21. reflectional symmetries over the lines connecting opposite vertices of the hexagon; reflectional symmetries over the lines connecting opposite midpoints of the sides of the hexagon; rotational symmetry with a 120° angle of rotation

Connections

In 19–21, describe in detail the symmetries in the design.

19.

20.

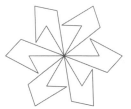

21.

22. The diagram below has rotational symmetry.

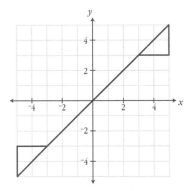

a. What center and angle of rotation will rotate one flag onto the other?

b. Compare the coordinates of key points on one flag with the coordinates of the images of those points on the other flag. Describe the pattern you see.

23. In this diagram, a 90° rotation about the origin will match one flag with another flag.

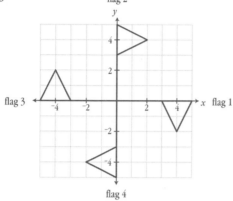

a. Under a 90° rotation about the origin, point (5, 0) on flag 1 is moved to point (0, 5) on flag 2. You can express this by writing (5, 0) → (0, 5). Write similar statements to show how the rotation moves each of the key points in flag 1 to a key point in flag 2.

b. Write statements to show how the rotation moves each key point in flag 2 to a key point in flag 3.

c. Write statements to show how the rotation moves each key point in flag 3 to a key point in flag 4.

d. Write statements to show how the rotation moves each key point in flag 4 to a key point in flag 1.

e. Use your results to write a general rule for finding the image of a point (*x*, *y*) under a 90° rotation about the origin.

22a. center at (0, 0); 180° angle of rotation

22b. Each coordinate of an image point is the opposite of the corresponding coordinate of the original point.

23a. (5, 0) → (0, 5); (4, ⁻2) → (2, 4); (3, 0) → (0, 3)

23b. (0, 5) → (⁻5, 0); (2, 4) → (⁻4, 2); (0, 3) → (⁻3, 0)

23c. (⁻5, 0) → (0, ⁻5); (⁻4, 2) → (⁻2, ⁻4); (⁻3, 0) → (0, ⁻3)

23d. (0, ⁻5) → (5, 0); (⁻2, ⁻4) → (4, ⁻2); (0, ⁻3) → (3, 0)

23e. (*x*, *y*) → (⁻*y*, *x*)

24. One possible design is shown below right. It was created by reflecting the original over the *y*-axis and then reflecting both the original and its image over the *x*-axis.

25. One possible design is shown below right. It was created by rotating the original 90° about the origin, and then rotating the image 90° about the origin, and finally rotating the second image 90° about the origin.

Extensions

24. The figure below is reproduced on Labsheet 3.ACE2. Use symmetry transformations to create a final design that has at least two lines of symmetry. Describe the transformations you used and the order in which you applied them.

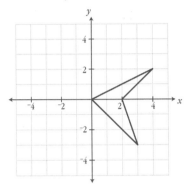

25. The figure below is reproduced on Labsheet 3.ACE2. Use symmetry transformations to create a final design that has rotational symmetry. Describe the transformations you used and the order in which you applied them.

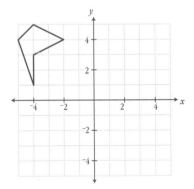

24.

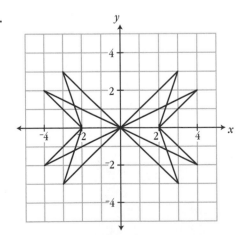

25.

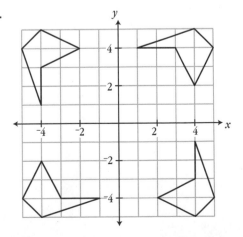

26. The figure below is reproduced on Labsheet 3.ACE2. Use symmetry transformations to create a final design that has both reflectional and rotational symmetry. Describe the transformations you used and the order in which you applied them.

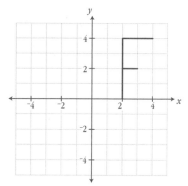

27. Investigate what happens when you rotate a figure 180° about a point and then rotate the image 180° about a different point. Is the combination of the two rotations equivalent to a single transformation? Test several cases, and make a conjecture about the result. You might start your investigation by rotating each polygon below 180° about C_1 and then 180° about C_2. These figures are reproduced on Labsheet 3.ACE3.

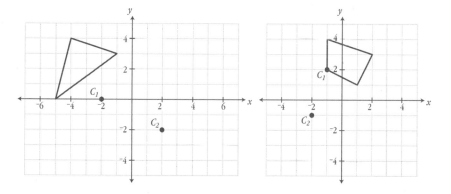

---

26. One possible design is shown below left. It was created by first rotating the original 90°, 180°, and 270°. The entire drawing was then reflected, first over the *x*-axis and then over the *y*-axis. The final drawing has rotational symmetry with a 90° angle of rotation and reflectional symmetry over the *x*-axis, the *y*-axis, the line $y = x$, and the line $y = ^-x$.

27. See page 58n.

26.

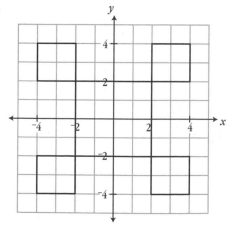

Possible Answers

1. $(^-x, y)$
2. $(x, ^-y)$
3. (y, x)
4. $(^-y, x)$
5. $(^-x, ^-y)$
6. $(y, ^-x)$
7. (x, y)
8. $(x + 2, y - 4)$
9. $(x - 6, y - 2)$

10. Two shapes are congruent if you can flip, slide, or turn one shape so it fits exactly on the other. This means that corresponding angles have the same measure and corresponding sides have the same length. You can determine whether two shapes are congruent by cutting out one shape and trying to match it with the other shape or by making measurements of sides and angles. Or, you can look for a series of transformations that will move one shape onto the other.

Mathematical Reflections

In this investigation, you looked at transformations of figures drawn on coordinate grids. You found that you could describe some symmetry transformations by telling what happens to a general point (x, y). These questions will help you summarize what you have learned:

1. What is the image of point (x, y) under a reflection over the y-axis?

2. What is the image of point (x, y) under a reflection over the x-axis?

3. What is the image of point (x, y) under a reflection over the line $y = x$?

4. What is the image of point (x, y) under a 90° rotation about the origin?

5. What is the image of point (x, y) under a 180° rotation about the origin?

6. What is the image of point (x, y) under a 270° rotation about the origin?

7. What is the image of point (x, y) under a 360° rotation about the origin?

8. What is the image of point (x, y) under the translation that slides (1, 2) to (3, $^-$2)?

9. What is the image of point (x, y) under a translation of 6 units to the left followed by a translation of 2 units down?

10. What does it mean for two shapes to be congruent? How can you determine whether two shapes are congruent?

Think about your answers to these questions, discuss your ideas with other students and your teacher, and then write a summary of your findings in your journal.

Tips for the Linguistically Diverse Classroom

Diagram Code The Diagram Code technique is described in detail in *Getting to Know Connected Mathematics*. Students use a minimal number of words and drawings, diagrams, or symbols to respond to questions that require writing. Example: Question 10—A student might respond to this question by drawing examples of congruent designs labeled *flip, slide,* and *turn* under the heading *Congruent*, and marking corresponding angles and sides of the congruent designs. Under the heading *Congruent?*, the student could draw a stick figure holding a pair of scissors, a piece of paper with a design on it, and the congruent "cutout" design.

TEACHING THE INVESTIGATION

3.1 • Writing Rules for Reflections

This problem introduces students to coordinate rules for transformations in the exploration of some simple cases.

Launch

Demonstrate the idea of creating designs by specifying the endpoints of line segments. Use the example in the student edition, which is reproduced on Transparency 3.1A, or create one of your own. The drawing instructions given here are generic, but they are much like those used in the TI series of graphing calculators and, though hidden from the user's view, in many computer drawing programs.

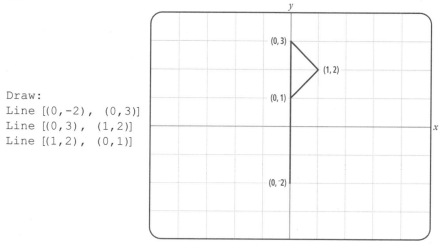

```
Draw:
Line [(0,-2), (0,3)]
Line [(0,3), (1,2)]
Line [(1,2), (0,1)]
```

Verify that students see the relationship between the commands and the drawing they produce.

Help the class begin to analyze such figures by locating the points that are useful for determining what happens to the figure under a transformation. Four points, for example, essentially determine the flag shown above. Explain that directions are given to draw line segments between pairs of these points; the figure can be reproduced by another human being or by a computer.

Discuss the questions in the "Think about this!" feature:

■ Is there a different set of commands that would create the same flag?

■ What commands would create a square centered at the origin?

■ What commands would create a nonsquare rectangle?

Have students share the commands they write. Two other sets of commands that would create the same flag are the following:

```
Draw:
Line [(0,3), (1,2)]
Line [(0,-2), (0,3)]
Line [(1,2), (0,1)]
```

```
Draw:
Line [(0,1), (0,-2)]
Line [(0,1), (0,3)]
Line [(1,2), (0,3)]
Line [(1,2), (0,1)]
```

If directions are given to communicate what happens to these points under a desired transformation, the image of the figure can be drawn.

If we look closely at how these endpoints are transformed, it might help us to come up with a rule for drawing a figure under a particular transformation.

For the Teacher: Using Computers to Enhance the Activity

If possible, make a computer drawing program available to students. Demonstrate the use of the program, and ask students how they think the computer knows where to draw things. Exploring what information the computer needs to perform a symmetry transformation will increase students' interest as well as their understanding.

Assign the problem and the follow-up to be done in groups of two to four students. Distribute copies of Labsheet 3.1A or grid paper to the groups, and have grid paper and tracing paper available for students to use during the exploration. Groups will need Labsheet 3.1B for the follow-up.

Explore

To write the new sets of commands, students should draw and analyze the reflection image called for, focusing on the significant points in the image. They should realize that these are the same points that are significant in the original.

In part D, students may need assistance with drawing the line $y = x$.

As you circulate, ask students if they see any pattern in what happens to the endpoints of the line segments under a particular reflection. If any students have particularly creative or insightful observations, make a note to call on them during the summary.

Some students may need help in the follow-up writing the first rule for finding the image of any point (x, y). If so, ask such questions as the following:

What pattern do you see between the coordinates of each point and its image?

How does the x value of each point change? How does the y value of each point change?

Summarize

Give students a chance to talk about what they observed in each part of the problem and to describe any patterns they see in what happens to the coordinate pairs. Call on particular students you observed or on others to share their sets of commands for each part. To check students' understanding during the discussion, draw a triangle such as the following on a transparent grid or the board. After discussing each part of the problem, ask for a similar set of commands to draw the image of the triangle. This gives immediate practice and keeps every student engaged during the summary.

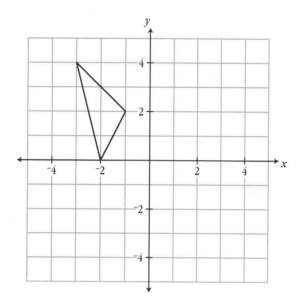

There are many sets of commands that would produce each figure. For example, the following two sets of commands would draw the image of the flag under a reflection over the *y*-axis:

```
Draw:                           Draw:
Line [(-5,5), (-5,3)]           Line [(0,0), (-3,3)]
Line [(-5,5), (0,0)]            Line [(-3,3), (-5,3)]
Line [(-5,3), (-3,3)]           Line [(-5,3), (-5,5)]
                                Line [(-5,5), (0,0)]
```

Move to the follow-up questions, and again have students share their answers to each part. The goal is for students to be able to describe the three reflections by writing a rule for finding the image of a general point (*x*, *y*).

> If you reflect a figure over the *x*-axis, how do the coordinates change? What rule could you write to represent this?
>
> What about over the *y*-axis? What rule could you write?
>
> How about over the line *y* = *x*?

To assess and extend students' understanding, you could ask them to write a set of commands that would draw the image of the flag under a reflection over a different line, such as *y* = ⁻*x*, *x* = 2, *y* = 3, *x* = ⁻5, or *y* = ⁻2.

3.2 • Writing Rules for Translations

In this problem, students examine pairs of points and images for various translations. The objective is for students to look for regular features of algebraic representations of translations so that they can both produce rules for translations and interpret such rules. Coordinate rules for translations are important throughout mathematics; developing informal knowledge through experience gives students valuable time to build their intuitive understanding.

Launch

Direct students' attention to the two designs shown in the student edition and reproduced on Transparencies 3.2A and 3.2B.

> What type of symmetry do you see in these designs? *(translational symmetry)*
>
> How can you tell that each design exhibits translational symmetry?
>
> In this problem, you will look for patterns that relate each flag to the flag next to it.

Have students work in pairs on the problem and the follow-up.

Explore

If students are having difficulty, focus their attention on the pairs of coordinates between the original and the first image. Have them organize their work so that it is easy to observe patterns in how the coordinates of the vertices change under the translation.

The follow-up is a chance for students to experiment on their own. When they begin working, point out that the questions do not specify the *length* of the translation. The context is creating designs with translational symmetry; the magnitude of the translation is up to the students. Stress that students should be looking for patterns in what happens under the translations. Have grid paper available for students to create the patterns.

Some students may be ready to express the relationships in questions 1 and 2 by writing coordinates that include a variable that represents the length of the translation; others will describe the relationships in words. You can introduce the idea of using a variable when you summarize the follow-up.

Listen to students as you circulate, looking for those who should be called on in the summary and for concepts with which students are struggling. This should give you ideas about what to emphasize during the summary.

If a group seems ready, extend their understanding by asking how the slopes of the lines in question 3 are related to the rules for the translations.

> When you translate a figure along the line $y = \frac{1}{2}x$, how is the change in the *x*-coordinate related to the change in the *y*-coordinate? *(When the x-coordinate increases by 2, the y-coordinate increases by 1.)*

What is the slope of the line $y = \frac{1}{2}x$? $(\frac{1}{2})$ How does the slope of the line relate to translating a figure along that line? *(The slope tells us that the y-coordinate would change by $\frac{1}{2}$ times whatever the x-coordinate changes by.)*

Summarize

Call on students to share their answers for each part of the problem. Then ask them to share the specific patterns they observed in their answers to the follow-up questions, leading them to write general rules for the translations.

Give me a few rules showing what might happen to a point (x, y) under a translation parallel to the x-axis.

Students might suggest such rules as $(x, y) \rightarrow (x + 2, y)$, $(x, y) \rightarrow (x - 4, y)$, and $(x, y) \rightarrow (x + 0.5, y)$.

What do the rules for translations parallel to the x-axis have in common? *(The rules all add or subtract a number from the x-coordinate, but leave the y-coordinate alone.)* Can you write a "super rule" to handle all of these cases?

Help students to see that a general rule is $(x, y) \rightarrow (x + a, y)$, where a gives the length and direction of the change parallel to the x-axis. If a is positive, the slide is to the right. If a is negative, the slide is to the left.

Now give me a few rules showing what might happen to a point (x, y) under a translation parallel to the y-axis.

Students might suggest such rules as $(x, y) \rightarrow (x, y + 1)$, $(x, y) \rightarrow (x, y - 4)$, and $(x, y) \rightarrow (x, y + 0.5)$.

What do the rules for translations parallel to the y-axis have in common? *[These rules are of the form $(x, y) \rightarrow (x, y + a)$, where a gives the length and direction of the change parallel to the y-axis.]*

Ask students to give you the rules for translations along each line in question 3 as you record them on the board.

Now give me a few rules showing what might happen to a point (x, y) under a translation parallel to the line $y = x$.

Students might suggest such rules as $(x, y) \rightarrow (x + 1, y + 1)$, $(x, y) \rightarrow (x - 4, y - 4)$, and $(x, y) \rightarrow (x + 0.5, y + 0.5)$.

What do the rules have in common? *[They are of the form $(x, y) \rightarrow (x + a, y + a)$, where a gives the length and direction of the change in the x direction, which is identical to the change in the y direction.]*

Give me a few rules showing what might happen to a point (x, y) under a translation parallel to the line $y = {}^-x$.

Some rules are $(x, y) \rightarrow (x + 1, y - 1)$, $(x, y) \rightarrow (x - 4, y + 4)$, and $(x, y) \rightarrow (x + 0.5, y - 0.5)$.

What do the rules have in common? *[They are of the form $(x, y) \rightarrow$ (x + a, y – a), where a gives the length and direction of the change in the x direction; the change in the y direction is of the same length but in the opposite direction.]*

And now give me a few rules showing what might happen to a point (x, y) under a translation parallel to the line $y = \frac{1}{2}x$.

Some rules are $(x, y) \rightarrow (x + 2, y + 1)$, $(x, y) \rightarrow (x - 4, y - 2)$, and $(x, y) \rightarrow (x + 1, y + 0.5)$.

What do the rules have in common? *[They are of the form $(x, y) \rightarrow$ (x + a, y + $\frac{1}{2}$a), where a gives the length and direction of the change in the x direction; the change in the y direction is half the length of the change in the x direction.]*

Help students see the relationship between the slope of the line along which the translation is taking place and the rules for the transformations.

How does the slope of the line $y = \frac{1}{2}x$ relate to the rules for translating a figure along that line? *(The slope $\frac{1}{2}$ tells us that the y-coordinate would change by $\frac{1}{2}$ of whatever the x-coordinate changes by.)*

Consider this rule for a translation: $(x, y) \rightarrow (x - 3, y + 2)$. Along what line does this translation occur? *(A change of negative 3 in the x direction and positive 2 in the y direction means that the slope of the line is $-\frac{2}{3}$. Therefore, the line is $y = -\frac{2}{3}x$.)*

Questions 3, 4, and 5 should lead students to a method for testing whether a given rule specifies a translation.

If you are given a rule, how can you recognize whether it describes a translation? *(It would show that something is added to or subtracted from x or y or both.)*

What about the rule $(x, y) \rightarrow (2x, y)$? Does it describe a translation?

You might sketch an example of a figure being transformed according to this rule. The square shown at the top of the next page increases in size in the x direction, which means this rule does not describe a translation.

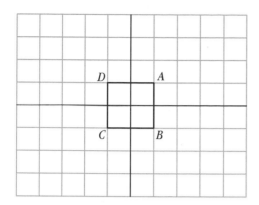 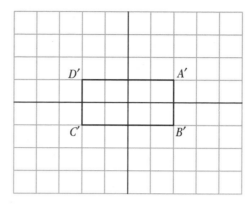

To assess students' understanding, write the following rules on the board and ask which represent translations:

$$(x, y) \rightarrow (x, y - 10) \qquad (x, y) \rightarrow (\tfrac{1}{2}x, y + \tfrac{1}{2})$$

$$(x, y) \rightarrow (^-2 + x, y) \qquad (x, y) \rightarrow (x + 3, y - 5)$$

3.3 • Writing Rules for Rotations

This problem follows the format of the two earlier problems, but this time the transformation is rotation. Only rotations that are 90° or multiples of 90° are considered because students can easily observe what happens to the coordinates. In high school, students in transformational geometry courses will learn to specify other turns algebraically.

Launch

Since this problem is similar to Problems 3.1 and 3.2, students should be able to anticipate the goal of the exploration. You might introduce the activity by making two copies of the triangle in the problem on transparencies and rotating one on top of the other about the origin to demonstrate a 90° rotation, a 180° rotation, a 270° rotation, and a 360° rotation. Remind students that rotations in this unit are done in a counterclockwise direction. Read through the problem with students as you show the rotations to emphasize that their goal is to observe what happens to the vertices of the triangle under the transformation.

Let students work in pairs on the problem and the follow-up. Distribute copies of Labsheet 3.3 or grid paper, and have grid paper and tracing paper available for use during the exploration.

Explore

As you circulate, listen for ideas that you would like shared in the summary. Also listen for ideas with which students are having trouble, and make a note to emphasize these in the summary.

Some students may need assistance with follow-up questions 3 and 4. To help them see that they should write out results of the reflections in terms of their effects on the coordinates, you might suggest that they record the original and final coordinates of triangle ABC and compare the results with the table from question 1. Students should then be able to see which of the rotations they have already done produces that same final image.

Summarize

As students share their answers to parts A through E, help the class to organize the data into a table as called for in follow-up question 1. With this evidence displayed, discussion of follow-up questions 2, 3, and 4 can help students make the leap to generalizing the result of a specific rotation.

Let's look carefully at each rotation, 90°, 180°, 270°, and 360°. For 90°, look at the points after each transformation. What happens to the coordinates? *(The x- and y-coordinates change places and then the sign of x changes.)*

Help the class to see why this makes sense.

If you were to rotate a grid 90°, what would happen to the *x*-axis? *(It would become the y-axis.)* What would happen to the *y*-axis? *(It would become the x-axis, but the positive end of the y-axis is now negative and the negative end is positive.)*

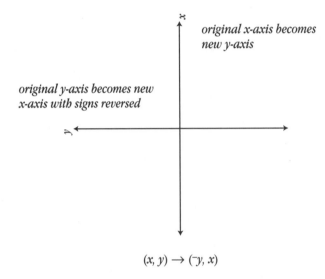

$(x, y) \rightarrow (^-y, x)$

What about a 180° turn? *(The signs of the coordinates are changed.)* Let's rotate the grid for a 180° turn.

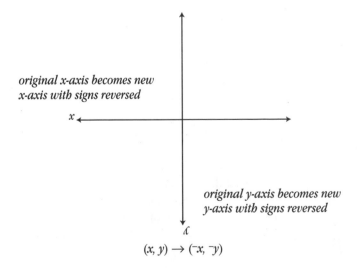

$(x, y) \rightarrow (^-x, ^-y)$

What about a 270° turn? *(The coordinates change places, and then the sign of y changes.)* Let's track the axes to see whether this makes sense.

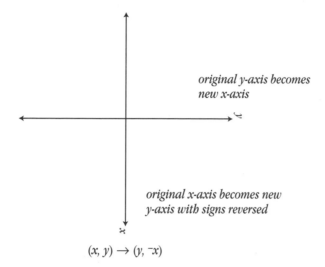

original y-axis becomes new x-axis

original x-axis becomes new y-axis with signs reversed

$(x, y) \rightarrow (y, {}^{-}x)$

Go over the answers to follow-up questions 3 and 4. Then ask:

How do these answers relate to what you found out earlier about reflecting a figure over two intersecting lines?

Help students see that the *y*-axis and the *x*-axis are intersecting lines, as are the *y*-axis and the line *y* = *x*. In both cases, reflecting over the two lines is equivalent to a single rotation, as students discovered in the previous investigation for other pairs of intersecting lines.

3.4 • Relating Symmetry to Congruence

As the title suggests, this problem will help students to connect the ideas of symmetry and transformations and the familiar geometric concept of congruence. Students will look at congruence in a more sophisticated way, using their knowledge of symmetry transformations to determine whether two figures are congruent.

Launch

Fundamentally, two figures are congruent if they have the same size and the same shape. The most natural, intuitive test for congruence is to attempt to superimpose a copy of one shape on top of another. When the figures are polygons, this superposition can be efficiently described as a correspondence of vertices and the corresponding sides and angles. The problem and follow-up introduce these ideas. Review what students know by asking:

What does it mean for two figures to be congruent?

Then explain:

> We can also think of congruence in terms of reflections, translations, and rotations.

Read through the problem and the follow-up as a class. Explain that students will need to gather evidence by taking measurements to confirm their ideas.

Have students work individually on the problem and the follow-up. Alternatively, this problem could be assigned as an exploration to be done at home and discussed in class.

Labsheet 3.4 contains several copies of the pentagons. You could give each student a full labsheet, or cut the labsheets in half. Students will also need rulers and angle rulers or protractors.

Explore

As students finish, you might engage them in an activity in which they use transparent reflection tools, mirrors, tracing paper, and any other tools to draw two simple, congruent figures. Students will need time to make accurate drawings and to copy the figures on clean paper showing none of their work lines. Then students can challenge a partner to try to get from one drawing to the other with a series of transformations.

For the Teacher: Equality versus Congruence

There is a natural tendency for students, teachers, and even mathematicians to be somewhat careless about the difference between equality and congruence. If your students are ready, you might discuss this distinction with them, though informality at this stage is fine.

The convention is to say that two line segments or angles are *equal* if they consist of the same sets of points—in other words, if they are identical. If two line segments have the same length, the segments are said to be *congruent* while the lengths are said to be *equal*. If two angles have the same measure, the angles are said to be *congruent* while the angle measures are said to be *equal*.

There are notation conventions for these concepts as well. A segment is often indicated as a pair of points with a bar above the endpoint labels; for example, $\overline{AB}$. That same pair of endpoint labels without the bar often represents the length of the segment. Similarly, when an angle sign is preceded by a lowercase "m," it is customary to understand that it is the *measure of the angle,* not the set of points, that is intended. Equality is indicated with the sign "="; congruence is indicated with the sign "≅." For example:

$\overline{AB} \cong \overline{CD}$ Segment *AB* is congruent to segment *CD*.

$AB = CD$ The length of segment *AB* equals the length of segment *CD*.

$\angle A \cong \angle T$ Angle *A* is congruent to angle *T*.

$m\angle A = m\angle T$ The measure of angle *A* equals the measure of angle *T*.

Summarize

Discuss the problem, concentrating particularly on part D. Have students demonstrate various combinations of transformations that would move pentagon *ABCDE* onto pentagon *PQRST*.

Additional Answers

Answers to Problem 3.1 Follow-Up

1. *A* (⁻3, 3), *B* (⁻2, 4), *C* (3, 3), *D* (1, 2), *E* (4, 1), *F* (2, ⁻2), *G* (3, ⁻4), *H* (5, ⁻5), *I* (⁻2, ⁻4), *J* (⁻4, ⁻1)

2. a.

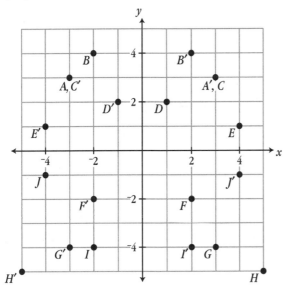

   b. *A′* (3, 3), *B′* (2, 4), *C′* (⁻3, 3), *D′* (⁻1, 2), *E′* (⁻4, 1), *F′* (⁻2, ⁻2), *G′* (⁻3, ⁻4), *H′* (⁻5, ⁻5), *I′* (2, ⁻4), *J′* (4, ⁻1)

3. a.

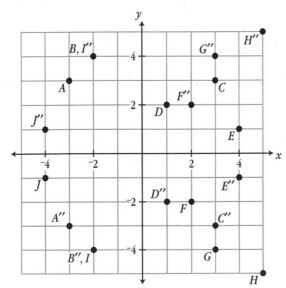

   b. *A″* (⁻3, ⁻3), *B″* (⁻2, ⁻4), *C″* (3, ⁻3), *D″* (1, ⁻2), *E″* (4, ⁻1), *F″* (2, 2), *G″* (3, 4), *H″* (5, 5), *I″* (⁻2, 4), *J″* (⁻4, 1)

4. a.

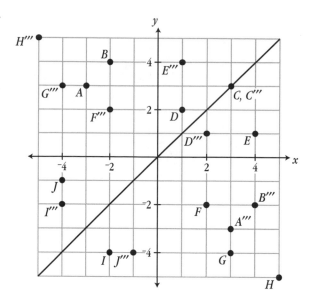

b. *A′″* (3, ⁻3), *B′″* (4, ⁻2), *C′″* (3, 3), *D′″* (2, 1), *E′″* (1, 4), *F′″* (⁻2, 2), *G′″* (⁻4, 3), *H′″* (⁻5, 5), *I′″* (⁻4, ⁻2), *J′″* (⁻1, ⁻4)

Answers to Problem 3.2

A. 1. *flag 2*
```
Draw:
Line [(-3,-4), (-3,2)]
Line [(-3,2), (-2,1)]
Line [(-2,1), (-3,0)]
```

flag 3
```
Draw:
Line [(-1,-4), (-1,2)]
Line [(-1,2), (0,1)]
Line [(0,1), (-1,0)]
```

flag 4
```
Draw:
Line [(1,-4), (1,2)]
Line [(1,2), (2,1)]
Line [(2,1), (1,0)]
```

flag 5
```
Draw:
Line [(3,-4), (3,2)]
Line [(3,2), (4,1)]
Line [(4,1), (3,0)]
```

B. 1. *flag 1*
```
Draw:
Line [(-4,-4), (-4,0)]
Line [(-4,0), (-3,-1)]
Line [(-3,-1), (-4,-2)]
```

flag 2
```
Draw:
Line [(-2,-3), (-2,1)]
Line [(-2,1), (-1,0)]
Line [(-1,0), (-2,-1)]
```

flag 3
```
Draw:
Line [(0,-2), (0,2)]
Line [(0,2), (1,1)]
Line [(1,1), (0,0)]
```

flag 4
```
Draw:
Line [(2,-1), (2,3)]
Line [(2,3), (3,2)]
Line [(3,2), (2,1)]
```

flag 5
```
Draw:
Line [(4,0), (4,4)]
Line [(4,4), (5,3)]
Line [(5,3), (4,2)]
```

Answers to Problem 3.2 Follow-Up

2. a. The y-coordinate of the figure directly above a given figure is y plus the length of the translation; the x-coordinate does not change.

    b. The y-coordinate of the second figure above a given figure is the y-coordinate of the first copy plus the length of the translation. It is also the y-coordinate of the original figure plus twice the length of the translation. In both cases, the x-coordinate does not change.

    c. The y-coordinate of any figure below a given figure is y minus the length of the translation; the x-coordinate does not change.

3. Answers will depend on the designs students draw; the length of the translation is not specified in the problem statement. *Possible translations along the line $y = x$:* $(x, y) \rightarrow (x + 1, y + 1) \rightarrow (x + 2, y + 2) \rightarrow (x + 3, y + 3)$, and so on. *Possible translations along the line $y = {}^-x$:* $(x, y) \rightarrow (x - 1, y + 1) \rightarrow (x - 2, y + 2) \rightarrow (x - 3, y + 3)$, and so on. *Possible translations along the line $y = \frac{1}{2}x$:* $(x, y) \rightarrow (x + 2, y + 1) \rightarrow (x + 4, y + 2) \rightarrow (x + 6, y + 3)$, and so on.

4. The rules are similar in that something is added to or subtracted from the x-coordinate and something is added to or subtracted from the y-coordinate. (**Teaching Tip:** You may want to explore with students the fact that for a constant change in the x-coordinate, there is a corresponding constant change in the y-coordinate. In fact, the ratio of the change in y to the change in x is the slope of the line used for the symmetry translation. For example, using the line $y = \frac{1}{2}x$, which has a slope of $\frac{1}{2}$, the change in y is 1 while the change in x is 2. This connects to and extends the concepts presented in the grade 7 unit on linear relationships, *Moving Straight Ahead.*)

For the Teacher: A Connection to Parabolas

In the unit *Frogs, Fleas, and Painted Cubes,* you may have had students explore changes in graphs of parabolas in ways that can now be connected to the ideas of algebraic representations of symmetry transformations. For example, the graph of $y = x^2 + 2$ is a simple translation of the graph of $y = x^2$. The graph is shifted up by two units, or the y values are increased by 2.

---

5. The rule $(x, y) \rightarrow (x + 2, y - 3)$ describes a translation that moves each point 2 units to the right and 3 units down. The rule $(x, y) \rightarrow (x - 2, y + 1)$ describes a translation that moves each point 2 units to the left and 1 unit up.

Answers to Problem 3.3 Follow-Up

3. A reflection over the y-axis reverses the sign of the x-coordinate, and a reflection over the x-axis reverses the sign of the y-coordinate. The two reflections do the following: $(x, y) \rightarrow ({}^-x, y) \rightarrow ({}^-x, {}^-y)$. This is the same result as a rotation of 180°.

4. A reflection over the y-axis reverses the sign of the x-coordinate, and a reflection over the line $y = x$ reverses the signs of both coordinates. The two reflections do the following: $(x, y) \rightarrow ({}^-x, y) \rightarrow (y, {}^-x)$. This is the same result as a rotation of 270°.

Answers to Problem 3.4 Follow-Up

1. a. *A* and *C*, *B* and *D*, *D* and *B*

   b. a rotation of 180° about the midpoint of segment *BD*

   c. sides: *AB* and *CD*, *DA* and *BC*, *BD* and *DB*; angles: ∠*A* and ∠*C*, ∠*ABD* and ∠*CDB*, ∠*ADB* and ∠*CBD*

2. a. You could measure to determine whether their diameters (or radii) are equal.

   b. You could translate one circle onto the other along a line joining their centers; rotate one onto the other, with the center of rotation being the midpoint of the line segment joining their centers; or reflect one onto the other over a line perpendicular to the line segment joining their centers and passing through the midpoint of that segment.

ACE Answers

Extensions

27. A 180° rotation about one point followed by a 180° rotation about a second point is equivalent to a translation parallel to the line connecting the two points. In each example given, the original polygon can be translated to match image 2.

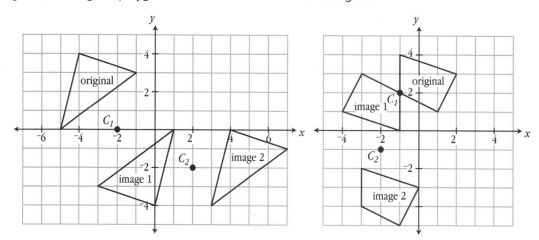

Symmetry and Algebra

In this investigation, students explore combinations of symmetry transformations for geometric figures and determine whether the operation of combining symmetry transformations satisfies some important algebraic properties.

In Problem 4.1, Transforming Triangles, students work with an equilateral triangle, which has six symmetries, analyzing what single transformation produces the same result as, or is equivalent to, each possible combination of two transformations. The process of performing one transformation on a figure followed by a second transformation on its image is referred to as the "and then" operation. Students produce and inspect an operation table showing the results of the "and then" operation and compare that table with a similar table for the arithmetic operation of multiplication. In Problem 4.2, Transforming Squares, they follow the same procedure, analyzing what single transformation is equivalent to each possible combination of two symmetry transformations for a square. In Problem 4.3, Properties of the Combining Operation, students review three important properties of addition and multiplication of real numbers—the commutative property, the identity property, and the inverse property. They are then asked to analyze whether the operation they have been studying—the "and then" operation of combining symmetry transformations—satisfies some or all of these properties.

Mathematical and Problem-Solving Goals

- *To determine the possible symmetry transformations for a given polygon*

- *To construct a table showing all possible results of combining two symmetry transformations of a given polygon*

- *To analyze such a table to determine whether (1) there is an identity element for the "and then" operation, (2) each element has an inverse for the "and then" operation, and (3) the "and then" operation is commutative*

Materials		
Problem	**For students**	**For the teacher**
All	Graphing calculators	Transparencies: 4.1 to 4.3B (optional), overhead graphing calculator (optional)
4.1	Paper copies of triangle *ABC* (1 per student; see blackline master), Labsheets 4.1A and 4.1B (1 each per student)	Transparencies of Labsheets 4.1A and 4.1B (optional)
4.2	Paper copies of square *ABCD* (1 per student; see blackline master), Labsheets 4.2A and 4.2B (1 each per student)	Transparencies of Labsheets 4.2A and 4.2B (optional)
ACE	Labsheet 4.ACE (1 per student)	Transparency of Labsheet 4.ACE (optional)

Symmetry and Algebra

A geometric figure is *symmetric* if a reflection or a rotation of the figure produces an image that matches the original figure exactly. Some figures have several symmetries. For example, an equilateral triangle has three lines of symmetry and can be rotated 120°, 240°, or 360° to exactly match the original figure.

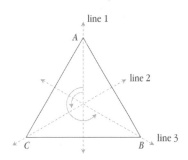

Your teacher will give you a copy of triangle *ABC* and the drawing above. Copy each vertex label onto the back of the triangle.

You can use your triangle as a tool for studying combinations of symmetry transformations. The drawing showing triangle *ABC* and its lines of symmetry will serve as a "frame" to help you keep track of the transformations. To begin, place your triangle in the frame, matching the vertices on your triangle with the vertices on the frame. To model a reflection over line 1 followed by a reflection over line 2, flip your triangle first over line 1 and then over line 2.

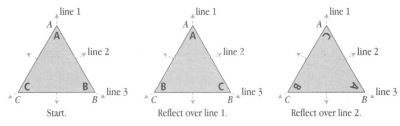

What single transformation would produce the same result as the combination of the two reflections?

Now use your triangle to find the result of a reflection over line 3 followed by a rotation of 120°. Is there a single transformation that would give the same result?

In this investigation, you will explore combinations of symmetry transformations for equilateral triangles and squares, and you will determine whether the operation of combining transformations satisfies some important algebraic properties.

Investigation 4: Symmetry and Algebra 59

4.1

Transforming Triangles

Grouping:
small groups

Launch

- As a class, explore and record the result of each transformation of an equilateral triangle.

- Introduce the idea of combinations of transformations.

- Have groups of two to four explore the problem and follow-up.

Explore

- Have students work in their groups but fill in their own tables.

Summarize

- Discuss the patterns in the "and then" table and what they reveal.

- Compare the "and then" and multiplication tables.

4.1 Transforming Triangles

The operations of arithmetic—addition, subtraction, multiplication, and division—define ways of putting two numbers together to get a single number. In a similar way, you can think of combining symmetry transformations as an operation that puts two transformations together to produce a single, equivalent transformation.

In this problem, you will explore combinations of symmetry transformations for an equilateral triangle. The notation L_n means a reflection over line n, and the notation R_n means a counterclockwise rotation of n degrees. The symbol $*$ represents the combining operation. You can read this symbol as "and then." For example, $L_1 * L_2 = R_{240}$ means that reflecting the triangle over line 1 "and then" reflecting it over line 2 is equivalent to rotating the triangle 240°.

> ### Problem 4.1
>
> The operation table below is reproduced on Labsheet 4.1B. Complete the table to show the results of combining symmetry transformations of an equilateral triangle. Each entry should be the result of performing the transformation in the left column followed by the transformation in the top row. The two entries already in the table represent the combinations you explored in the introduction:
>
> $$L_1 * L_2 = R_{240} \qquad L_3 * R_{120} = L_2$$
>
> | $*$ | R_{360} | R_{120} | R_{240} | L_1 | L_2 | L_3 |
> |---|---|---|---|---|---|---|
> | R_{360} | | | | | | |
> | R_{120} | | | | | | |
> | R_{240} | | | | | | |
> | L_1 | | | | | R_{240} | |
> | L_2 | | | | | | |
> | L_3 | | L_2 | | | | |
>
> Note that transformation R_{360}, a 360° rotation, carries every point back to where it started. As you combine transformations, you will discover that many combinations are equivalent to R_{360}.

 Problem 4.1 Follow-Up

1. Look carefully at the entries in your table. Describe any interesting patterns in the rows, columns, or blocks of entries. What do these patterns tell you about the results of combining rotations and line reflections?

Assignment Choices

ACE questions 7, 8, and unassigned choices from earlier problems

Answer to Problem 4.1

$*$	R_{360}	R_{120}	R_{240}	L_1	L_2	L_3
R_{360}	R_{360}	R_{120}	R_{240}	L_1	L_2	L_3
R_{120}	R_{120}	R_{240}	R_{360}	L_2	L_3	L_1
R_{240}	R_{240}	R_{360}	R_{120}	L_3	L_1	L_2
L_1	L_1	L_3	L_2	R_{360}	R_{240}	R_{120}
L_2	L_2	L_1	L_3	R_{120}	R_{360}	R_{240}
L_3	L_3	L_2	L_1	R_{240}	R_{120}	R_{360}

Answers to Problem 4.1 Follow-Up

See page 70l.

2. Make an operation table for multiplication of the whole numbers 1, 2, 3, 4, 5, and 6. Compare the patterns in your multiplication table with the patterns in your table of transformation combinations. Describe any interesting similarities and differences you discover.

4.2 Transforming Squares

If a polygon has reflectional or rotational symmetry, you can make an operation table to show the results of combining symmetry transformations for that polygon. There are eight symmetry transformations for a square, including a rotation of 360°. In this problem, you will investigate combinations of these symmetry transformations.

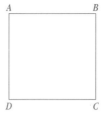

Problem 4.2

A. On Labsheet 4.2B, draw all the lines of symmetry and describe all the rotations that produce images that exactly match the original square. Label the lines of symmetry in clockwise order as line 1, line 2, and so on.

B. Cut out a copy of the square, and copy each vertex label onto the back of the square. Use this copy to explore combinations of symmetry transformations. The operation table below is reproduced on Labsheet 4.2B. Complete the table to show the results of combining pairs of transformations. When you enter the column heads, first list the rotations and then the reflections. Use the same order for the row heads.

$*$	R_{360}							
R_{360}								

Transforming Squares

Grouping:
small groups

Launch

- Have students explore and record the result of each transformation of a square.

- Have groups of two to four explore the problem and follow-up.

Explore

- Have students work in their groups but fill in their own tables.

Summarize

- Discuss the patterns in the table for the square and compare them with those for an equilateral triangle.

- Compare the "and then" and multiplication tables.

Answers to Problem 4.2

A. See page 70l.

B.

$*$	R_{360}	R_{90}	R_{180}	R_{270}	L_1	L_2	L_3	L_4
R_{360}	R_{360}	R_{90}	R_{180}	R_{270}	L_1	L_2	L_3	L_4
R_{90}	R_{90}	R_{180}	R_{270}	R_{360}	L_2	L_3	L_4	L_1
R_{180}	R_{180}	R_{270}	R_{360}	R_{90}	L_3	L_4	L_1	L_2
R_{270}	R_{270}	R_{360}	R_{90}	R_{180}	L_4	L_1	L_2	L_3
L_1	L_1	L_4	L_3	L_2	R_{360}	R_{270}	R_{180}	R_{90}
L_2	L_2	L_1	L_4	L_3	R_{90}	R_{360}	R_{270}	R_{180}
L_3	L_3	L_2	L_1	L_4	R_{180}	R_{90}	R_{360}	R_{270}
L_4	L_4	L_3	L_2	L_1	R_{270}	R_{180}	R_{90}	R_{360}

Assignment Choices

ACE questions 1a, 1b, 2a, 2b, 3a, 3b, 6, 11a, 11b, and unassigned choices from earlier problems (Note: 1a and 1b must be assigned before 6)

Properties of the Combining Operation

Launch

- Discuss the commutative property, identities, and inverses.

- Tell students they will explore whether the "and then" operation has these properties.

Explore

- Have pairs do the problem and follow-up, perhaps using a think-pair-share strategy.

Summarize

- As a class, analyze the "and then" table for commutativity, an identity, and inverses.

- Again compare the "and then" and multiplication operations.

Assignment Choices

ACE questions 1c–e, 2c–e, 3c–e, 4, 5, 9, 10, 11c–e, 12, 13, and unassigned choices from earlier problems

---

■ Problem 4.2 Follow-Up

1. Compare the patterns in the operation table for the square with the patterns in the operation table for the equilateral triangle. In what ways are the tables similar? In what ways are they different?

2. Make an operation table for multiplication of the whole numbers 1, 2, 3, 4, 5, 6, 7, and 8. Compare the patterns in your multiplication table with the patterns in your table of transformation combinations for the square. Describe any interesting similarities and differences you discover.

4.3 Properties of the Combining Operation

The operations of addition and multiplication satisfy important properties that are useful for reasoning about expressions and equations.

- The order in which numbers are added or multiplied does not affect the result. This is called the commutative property. We say that addition and multiplication are *commutative operations.* In symbols, if a and b are real numbers, then

$$a + b = b + a \quad \text{and} \quad a \times b = b \times a.$$

- Adding 0 to a number has no effect. Multiplying a number by 1 has no effect. In symbols, if a is a real number, then

$$0 + a = a + 0 = a \quad \text{and} \quad 1 \times a = a \times 1 = a.$$

We call 0 and 1 *identity elements;* 0 is the additive identity, and 1 is the multiplicative identity.

- For any number a, the sum of a and ^-a is 0, the additive identity. For any nonzero number a, the product of a and $\frac{1}{a}$ is 1, the multiplicative identity.

$$a + {}^-a = {}^-a + a = 0 \quad \text{and} \quad a \times \tfrac{1}{a} = \tfrac{1}{a} \times a = 1$$

We call ^-a the *additive inverse* of a and $\frac{1}{a}$ the *multiplicative inverse* of a.

In this problem, you will determine whether the operation for combining symmetry transformations for an equilateral triangle satisfies similar properties.

Answers to Problem 4.2 Follow-Up

1. Possible answer: Each cell contains one of the possible symmetry transformations, which means that every combination of two transformations is equivalent to a single transformation. The upper-left and lower-right sections are filled with rotations, and the lower-left and upper-right sections are filled with reflections. Specifically, combining a rotation and a reflection is equivalent to a single reflection; combining two rotations or two reflections is equivalent to a single rotation. Also, there is a row and a column that contain entries that match the row heads and column heads. The tables are different only in that the table for the square contains more symmetry transformations and thus more rows and columns.

2. See page 70m.

Problem 4.3

Refer to your operation table for combining symmetry transformations for an equilateral triangle.

A. Is $*$ a commutative operation? In other words, does the order in which you combine symmetry transformations make a difference? Justify your answer.

B. Like addition and multiplication, the $*$ operation has an identity element. That is, there is a transformation that has no effect when it is applied before or after another transformation. Tell which transformation is the identity element, and explain how you know.

C. Does each symmetry transformation have an inverse? That is, can you combine each transformation with another transformation, in either order, to get the identity element you found in part B? If so, list each transformation and its inverse.

■ **Problem 4.3 Follow-Up**

Refer to your operation table for combining symmetry transformations for a square.

1. Is $*$ a commutative operation for the symmetry transformations for a square?

2. What is the identity element in this situation?

3. Does each symmetry transformation have an inverse? If so, list each transformation and its inverse.

Answers to Problem 4.3

A.   The operation $*$ is not commutative for an equilateral triangle; order does matter in many cases. For example, reflecting the triangle over line 1 and then over line 2 is equivalent to rotating it 240°; reflecting it over line 2 and then over line 1 is equivalent to rotating it 120°.

B.   The symmetry transformation R_{360} is an identity element. Rotating the triangle 360° is equivalent to doing nothing because the triangle ends in its original position.

C.   Each line reflection and the rotation R_{360} can be combined with R_{360} to get the identity element, R_{360}. The remaining reflections, R_{120} and R_{240}, are inverses of each other since $R_{120} * R_{240} = R_{240} * R_{120} = R_{360}$.

Answers to Problem 4.3 Follow-Up

See page 70m.

ACE

Answers

Applications

1a, b. See below right.

1c. The $*$ operation is commutative for this set of symmetry transformations. The table shows that $a * b = b * a$ for every combination of transformations.

1d. R_{360}

1e. Each transformation is its own inverse:
$R_{360} * R_{360} = R_{360}$,
$R_{180} * R_{180} = R_{360}$,
$L_1 * L_1 = R_{360}$,
$L_2 * L_2 = R_{360}$.

2a. The triangle can be reflected over line 1, which passes through vertex A and the midpoint of side BC. It can be rotated $360°$.

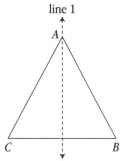

2b.

$*$	R_{360}	L_1
R_{360}	R_{360}	L_1
L_1	L_1	R_{360}

2c. The $*$ operation is commutative for this set of symmetry transformations. The table shows that $a * b = b * a$ for every combination of transformations.

2d. R_{360}

2e. Each transformation is its own inverse.

As you work on these ACE questions, use your calculator whenever you need it.

Applications

1. Rectangle $ABCD$ is not a square.

   a. Describe all the symmetry transformations for the rectangle.

   b. Make a table showing the results of combining pairs of symmetry transformations with the $*$ operation.

   c. Is $*$ a commutative operation for this set of symmetry transformations? Explain.

   d. What is the identity element in this situation?

   e. Match each symmetry transformation with its inverse.

2. Triangle ABC is isosceles with congruent sides AC and AB.

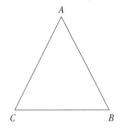

   a. Describe all the symmetry transformations for the triangle.

   b. Make a table showing the results of combining pairs of symmetry transformations with the $*$ operation.

   c. Is $*$ a commutative operation for this set of symmetry transformations? Explain.

   d. What is the identity element in this situation?

   e. Match each symmetry transformation with its inverse.

1a. The rectangle can be reflected over line 1, the vertical line through the center, and over line 2, the horizontal line through the center. It can be rotated $180°$ and $360°$ about the point where the lines of reflection intersect.

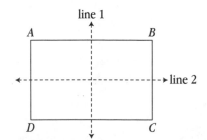

1b.

$*$	R_{360}	R_{180}	L_1	L_2
R_{360}	R_{360}	R_{180}	L_1	L_2
R_{180}	R_{180}	R_{360}	L_2	L_1
L_1	L_1	L_2	R_{360}	R_{180}
L_2	L_2	L_1	R_{180}	R_{360}

3. Study this symmetric design, which is composed of four birds.

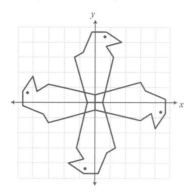

    a. Describe all the symmetry transformations for this design.

    b. Make a table showing the results of combining pairs of symmetry transformations with the ✳ operation.

    c. Is ✳ a commutative operation for this set of symmetry transformations? Explain.

    d. What is the identity element in this situation?

    e. Match each symmetry transformation with its inverse.

Connections

4. **a.** What are the additive inverses of 2.3, $^-4$, $\frac{2}{3}$, and $-\frac{5}{4}$?

    b. What are the multiplicative inverses of 2, 5, 0.4, and $\frac{2}{3}$?

    c. Is subtraction of real numbers a commutative operation? Justify your answer.

    d. Is division of real numbers a commutative operation? Justify your answer.

3a. The symmetry transformations for the four birds are R_{90}, R_{180}, R_{270}, and R_{360}.

3b. See below left.

3c. The ✳ operation is commutative for this set of symmetry transformations. The table shows that $a ✳ b = b ✳ a$ for every combination of transformations.

3d. R_{360}

3e. R_{360} is its own inverse; R_{180} is its own inverse; and R_{270} and R_{90} are inverses since $R_{270} ✳ R_{90} = R_{90} ✳ R_{270} = R_{360}$.

Connections

4a. The additive inverses are $^-2.3$, 4, $-\frac{2}{3}$, and $\frac{5}{4}$, respectively.

4b. The multiplicative inverses are $\frac{1}{2}$, $\frac{1}{5}$, 2.5, and $\frac{3}{2}$, respectively.

4c. no; $5 - 3 \neq 3 - 5$

4d. no; $2 \div 3 \neq 3 \div 2$

3b.

✳	R_{360}	R_{90}	R_{180}	R_{270}
R_{360}	R_{360}	R_{90}	R_{180}	R_{270}
R_{90}	R_{90}	R_{180}	R_{270}	R_{360}
R_{180}	R_{180}	R_{270}	R_{360}	R_{90}
R_{270}	R_{270}	R_{360}	R_{90}	R_{180}

5a. There is no identity element for subtraction. For a number to be an identity element, it must not have an effect when combined *in either order* with any other element; there is no such element for subtraction. For example, although 5 – 0 = 5, 0 – 5 ≠ 5, so 0 is not an identity element.

5b. There is no identity element for division. For example, although 5 ÷ 1 = 5, 1 ÷ 5 ≠ 5, so 1 is not an identity element.

6a. Combining two reflections is equivalent to performing a rotation. (Note: Since the lines of reflection for these shapes intersect, this is consistent with what students discovered in Problem 2.4 about reflecting a figure over intersecting lines.)

6b. Combining two rotations is equivalent to performing a single rotation.

6c. Combining a rotation and a reflection is equivalent to performing a line reflection.

7a. The x represents the transformation that is performed before R_{120} to produce a result equivalent to L_1. From the table, this must be L_2.

7b. The x represents the transformation that is performed after L_3 to produce a result equivalent to R_{240}. From the table, this must be L_1.

5. **a.** Is there an identity element for subtraction of real numbers? Explain.

b. Is there an identity element for division of real numbers? Explain.

6. Look at the operation tables for combining symmetry transformations for an equilateral triangle, a square, and a rectangle.

a. What do the patterns in the tables tell you about the result of combining two line reflections?

b. What do the patterns in the tables tell you about the result of combining two rotations?

c. What do the patterns in the tables tell you about the result of combining a rotation and a line reflection?

7. **a.** In the equation $x * R_{120} = L_1$, the x represents a symmetry transformation of an equilateral triangle. Use your table from Problem 4.1 to help you solve this equation. Explain how you found your answer. Check your solution by using a paper triangle to perform the transformations.

b. In the equation $L_3 * x = R_{240}$, the x represents a symmetry transformation of an equilateral triangle. Solve this equation, and explain how you found the solution. Check your solution by using a paper triangle to perform the transformations.

reflecting and rotating triangles

8. a. The figure below appears on Labsheet 4.ACE1. Sketch the image of the flag under the translation $(x, y) \rightarrow (x + 2, y + 3)$. We will call this translation T_1.

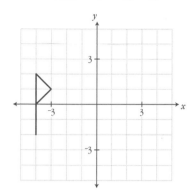

b. Sketch the image of the *original* flag under the translation $(x, y) \rightarrow (x + 4, y - 2)$. We will call this translation T_2.

c. Sketch the image of the original flag under the combination $T_1 * T_2$.

d. Sketch the image of the original flag under the combination $T_2 * T_1$.

9. a. Describe the relationship between pairs of opposite sides and between pairs of opposite angles for parallelogram $ABCD$.

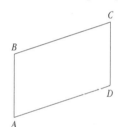

b. Suppose that translation T_2 matches point A with point B, and translation T_2 matches point A with point D.

  i. Where is the image of point A under the combination $T_1 * T_2$?

  ii. Where is the image of point A under the combination $T_2 * T_1$?

c. What conjecture would you make about commutativity of translations based on your answer to part b?

8. The images are shown below left. The image under $T_1 * T_2$ is the same as the image under $T_2 * T_1$.

9a. Opposite sides are parallel and have the same length; opposite angles have the same measure.

9b. i. The image of point A is located at point C.

ii. The image of point A is located at point C.

9c. The $*$ operation is commutative for this set of translations.

8.

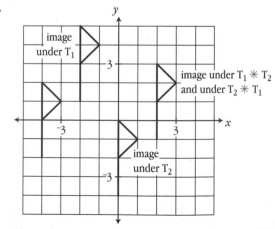

Extensions

10. See below right.

Teaching Tip: You may need to point out to students that since R_{360} is the identity element for symmetry transformations, any combination of a transformation with R_{360} is equivalent to that transformation.

11. See page 70m.

Extensions

10. Inverses are useful for solving algebraic equations. For example, to solve $x + 5 = 12$, you can add the additive inverse of 5 to both sides of the equation:

$$x + 5 = 12$$
$$x + 5 + {}^-5 = 12 + {}^-5$$
$$x = 7$$

To solve $4x = 36$, you can multiply both sides by the multiplicative inverse of 4:

$$4x = 36$$
$$\tfrac{1}{4}(4x) = \tfrac{1}{4}(36)$$
$$x = 9$$

Use a similar strategy to solve these equations involving symmetry transformations of a *square*. Explain the steps in your solution.

a. $x * R_{90} = R_{180}$

b. $L_1 * x = R_{180}$

c. $R_{270} * x = L_1$

11. a. Draw a regular pentagon. Describe all the symmetries in the pentagon.

b. Make a table showing the results of combining symmetry transformations with the $*$ operation.

c. Is the $*$ operation commutative for this set of symmetry transformations? Explain.

d. What is the identity element in this situation?

e. Match each symmetry transformation with its inverse.

10a.
$$x * R_{90} = R_{180}$$
$$x * (R_{90} * R_{270}) = R_{180} * R_{270}$$
$$x * R_{360} = R_{90}$$
$$x = R_{90}$$

10b.
$$L_1 * x = R_{180}$$
$$(L_1 * L_1) * x = L_1 * R_{180}$$
$$R_{360} * x = L_3$$
$$x = L_3$$

10c.
$$R_{270} * x = L_1$$
$$(R_{90} * R_{270}) * x = R_{90} * L_1$$
$$R_{360} * x = L_2$$
$$x = L_2$$

12. Suppose the symbol ◆ represents the operation of finding the mean. So, for the numbers a and b, $a ◆ b = \frac{a+b}{2}$.

a. Evaluate 12 ◆ 5 , 3 ◆ $^-$8 , and 2.5 ◆ 9.7.

b. Test a few examples to determine whether ◆ is a commutative operation.

c. Does this operation have an identity element?

13. Suppose the numbers from 1 to 7 represent the days of the week: 1 represents Monday, 2 represents Tuesday, and so on.

JANUARY 1997						
SUN	MON	TUE	WED	THU	FRI	SAT
			1	2	3	4
5	6	7	8	9	10	11
12	13	14	15	16	17	18
19	20	21	22	23	24	25
26	27	28	29	30	31	

a. What number represents the day 5 days after Tuesday? What number represents the day 6 days after Friday?

b. Suppose the expression $a ▲ b$ means the day b days after day a. For example, 5 ▲ 3 represents the day 3 days after Friday (day 5). Complete the table below.

▲	1	2	3	4	5	6	7
1	2	3					
2							
3							
4							
5							
6							
7							

c. Is ▲ a commutative operation? Explain.

d. What is the identity element for ▲?

e. Does each number have an inverse under the ▲ operation?

f. January 1, 1996, was a Monday. The year 1996 was a leap year, so it had 366 days. How can you show that January 1, 1997, occurred on a Wednesday?

Investigation 4: Symmetry and Algebra | **69**

13b.

▲	1	2	3	4	5	6	7
1	2	3	4	5	6	7	1
2	3	4	5	6	7	1	2
3	4	5	6	7	1	2	3
4	5	6	7	1	2	3	4
5	6	7	1	2	3	4	5
6	7	1	2	3	4	5	6
7	1	2	3	4	5	6	7

13f. If you divide 366 days by 7 days, you will find that there are 52 weeks plus 2 days in 366 days. Since January 1, 1996, was a Monday, 52 weeks later was another Monday. Two days beyond that was a Wednesday, so January 1, 1997, was a Wednesday. In terms of the ▲ operation, you can think about calculating 1 ▲ 7 ▲ 7 ▲ 7 ▲ $\cdots$ ▲ 7 ▲ 2. Since 7 is the identity element, 1 ▲ 7 ▲ 7 ▲ 7 ▲ $\cdots$ ▲ 7 ▲ 2 = 1 ▲ 2 = 3, which represents Wednesday.

12a. $12 ◆ 5 = \frac{12+5}{2} = \frac{17}{2}$ $= 8.5$; $3 ◆ {}^-8 = \frac{3 + {}^-8}{2} = \frac{-5}{2}$ $= {}^-2.5$; $2.5 ◆ 9.7 = \frac{2.5 + 9.7}{2}$ $= \frac{12.2}{2} = 6.1$

12b. The ◆ operation is commutative because the two numbers that are combined are added and then divided by 2. Since addition is commutative, the order in which the numbers are combined does not matter.

12c. There is not an identity element for the ◆ operation. Although $a ◆ a = a$ for any number a, there is no single element i such that $a ◆ i = i ◆ a = a$ for all a.

13a. Day 7 (Sunday) is 5 days after Tuesday. Day 4 (Thursday) is 6 days after Friday.

13b. See below left.

13c. The ▲ operation is commutative because $a ▲ b = b ▲ a$ for any days a and b. This can be seen in the symmetry about the diagonal in the table from upper left to lower right.

13d. The identity element is 7 because $7 ▲ a = a ▲ 7$ $= a$ for any day a.

13e. Each element a has an inverse such that a combined with its inverse, in either order, is equivalent to 7. The inverse pairs are 1 and 6, 2 and 5, 3 and 4, and 7 and 7.

13f. See below left.

Investigation 4 | **69**

1. In both tables, the entries in the first row match the column heads and the entries in the first column match the row heads. Every entry in the "and then" table is one of the transformations that appear in the row and column heads; the entries in the multiplication table consist of more numbers than just those in the row and column heads. The multiplication table has symmetry about the diagonal from upper left to lower right; that is not true for the "and then" table. This is because multiplication is commutative and the "and then" operation is not.

2a. A reflection followed by a reflection is equivalent to a rotation.

2b. A combination of two rotations is equivalent to another rotation.

2c. A rotation followed by a reflection or a reflection followed by a rotation is equivalent to a reflection.

3a. If the column and row heads are in the same order, you can check to see whether the table is symmetric about the diagonal from the upper left to the lower right. If so, the operation is commutative.

3c. Check to see that the identity element appears exactly once in each row and each column. If so, you can find pairs of inverse elements by finding the row and column heads for each identity entry.

Mathematical Reflections

In this investigation, you explored combinations of symmetry transformations for geometric shapes, and you determined whether the combining operation satisfies important algebraic properties. These questions will help you summarize what you have learned:

1 How are the patterns in the operation tables for combinations of symmetry transformations similar to the patterns in tables for the arithmetic operation multiplication? How are they different?

2 Explain what the operation tables show about the results of combining

   a. two line reflections.

   b. two rotations.

   c. a rotation and a line reflection.

3 Explain how you could use an operation table to check for

   a. commutativity.

   b. an identity element.

   c. inverse elements.

Think about your answers to these questions, discuss your ideas with other students and your teacher, and then write a summary of your findings in your journal.

3b. Check to see whether there is a row whose entries are identical to the column heads and a column whose entries are identical to the row heads. If so, and the head for that row is the same as the head for that column, that head is the identity element.

Tips for the Linguistically Diverse Classroom

Diagram Code The Diagram Code technique is described in detail in *Getting to Know Connected Mathematics*. Students use a minimal number of words and drawings, diagrams, or symbols to respond to questions that require writing. Example: Question 2a—A student might respond to this question by writing *reflection 1 * reflection 2 = rotation.*

TEACHING THE INVESTIGATION

4.1 • Transforming Triangles

In this problem, students will analyze all the symmetries of an equilateral triangle and explore the results of combinations of two symmetry transformations, trying to find a single transformation equivalent to each combination. They will organize their results into a table and look for patterns. They will also compare the table for the "and then" operation to a multiplication table.

Launch

To introduce the problem, draw an equilateral triangle on the board or the overhead.

> What kinds of symmetry does an equilateral triangle have? *(reflectional symmetry and rotational symmetry)* Where are the lines of symmetry?

Draw the lines of symmetry as students describe them, labeling them line 1, line 2, and line 3.

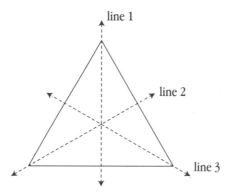

> Through what angles can an equilateral triangle be rotated about its center to look the same as the original? *(120°, 240°, and 360°)*

Distribute a paper copy of equilateral triangle *ABC* (cut from the blackline master in Additional Resources), Labsheet 4.1A, and Labsheet 4.1B to each student. Have students label the back side of each vertex of their paper triangle; this will help them to keep track of the position of each vertex under each transformation.

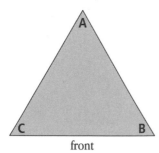
front

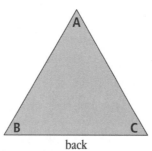
back

Lead the class in an exploration of the results of the six single transformations of an equilateral triangle, perhaps modeling each transformation at the overhead using a transparency of triangle *ABC* and a transparency of the triangle frame shown on Labsheet 4.1B. Students will record the results of each transformation on Labsheet 4.1A.

Let's see what happens when we reflect the triangle over line 1.

Place your paper triangle over the triangle frame shown at the top of Labsheet 4.1B. Match the vertices on your paper triangle with the vertices of the triangle frame. What happens when you reflect your triangle over line 1?

After the triangle is reflected over line 1, the triangle and frame will look as follows:

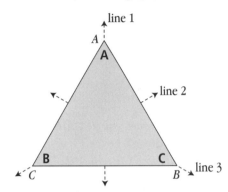

We can represent a reflection over line 1 with the notation L_1. Copy the results of this reflection onto Labsheet 4.1A into the triangle labeled L_1. Label the triangle to show what the vertices look like after the reflection.

Depending on the time available and students' understanding, continue exploring the transformations as a class or have students work in groups of two to four. Students should use their paper triangles to determine and sketch the results of the remaining five transformations: reflection about lines 2 and 3, and rotation about angles of 120°, 240°, and 360°. Explain that the notation R_{360} represents a counterclockwise rotation of 360°. The completed Labsheet 4.1A should look as follows:

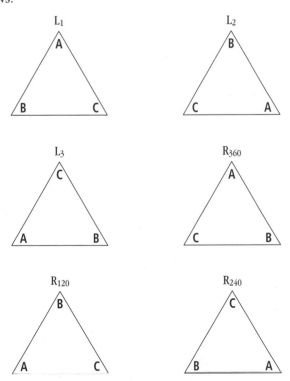

Now introduce the idea of performing a combination of two transformations.

> Use your triangle and the triangle frame to find the result of a reflection over line 1 and then over line 2. What is the result?

You might demonstrate this process at the overhead.

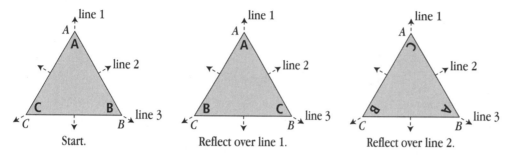

| Start. | Reflect over line 1. | Reflect over line 2. |

> Compare the final position of your paper triangle with the results of the single transformations shown on Labsheet 4.1A. What single transformation gives you the same result? *(a rotation of 240°, or R$_{240}$)*

Remind students to return their paper triangle to the original starting position by matching its labels to those on the triangle frame.

> Now use your triangle to find the result of a reflection over line 3 followed by a rotation of 120°. What single transformation gives the same result as the combination of these two transformations?
> *(a reflection over line 2, or L$_2$)*

> The operation of combining symmetry transformations is somewhat like operations of arithmetic, which define ways to combine two numbers. You know how to make tables showing the basic facts of addition, subtraction, multiplication, and division. Now you will make a table showing the results of combining symmetry transformations for an equilateral triangle.

Have students work in groups of two to four on the problem and the follow-up. Labsheet 4.1B contains the beginnings of the tables that students are to construct.

Explore

Students should work in their groups to determine the entries, with each student filling in his or her own table. You may want to distribute transparencies of Labsheet 4.1B to one or two groups for copying the tables from the problem and the follow-up for use in the class discussion.

Summarize

As a class, discuss the entries in the table for the combinations of symmetry transformations of an equilateral triangle, which is shown at the top of the next page. Resolve any differences by demonstrating the disputed entries and analyzing the results as a class.

$*$	R_{360}	R_{120}	R_{240}	L_1	L_2	L_3
R_{360}	R_{360}	R_{120}	R_{240}	L_1	L_2	L_3
R_{120}	R_{120}	R_{240}	R_{360}	L_2	L_3	L_1
R_{240}	R_{240}	R_{360}	R_{120}	L_3	L_1	L_2
L_1	L_1	L_3	L_2	R_{360}	R_{240}	R_{120}
L_2	L_2	L_1	L_3	R_{120}	R_{360}	R_{240}
L_3	L_3	L_2	L_1	R_{240}	R_{120}	R_{360}

What patterns do you see in the table? *(The upper-left and lower-right sections are filled with rotations. The other sections contain only reflections.)*

What does this tell you about the results of combining rotations and line reflections? *(It looks like combining two rotations or two reflections is equivalent to a single rotation and that combining a reflection and a rotation is equivalent to a single reflection.)*

What other patterns do you see? *(The first row contains the row heads as entries, and the first column contains the column heads as entries.)* And what is special about the first column and the first row? *(They include R_{360} as one of the transformations.)* How is the transformation R_{360} related to the fact that this row and column contain the same entries as the row and column heads? *(R_{360} just returns the triangle to its starting position, so it's like doing nothing. The result of a pair of combinations that involves R_{360} is equivalent to the result of the other transformation alone; these are the entries in this row and column.)*

What happens when you reflect the triangle over the same line twice? *(You get what you started with, which is equivalent to a rotation of 360°.)*

Help students connect the patterns in the table to their earlier explorations with combining transformations.

Notice that the lines of symmetry of an equilateral triangle are intersecting lines. So, when you reflect the triangle over one line of symmetry and then over a different line of symmetry, you are reflecting the triangle over *intersecting lines*. What do you recall about reflecting a figure over two *intersecting lines*? *(A combination of the two reflections is equivalent to a single rotation.)*

In your earlier work, you rotated a figure about a point and then rotated the image about the same point. What single transformation was equivalent to two rotations? *(A combination of two rotations is equivalent to a single rotation.)*

Does your table reflect these two results? *(yes)*

Ask students to compare the "and then" table with the multiplication table.

> In what ways are the "and then" operation on symmetry transformations and multiplication on numbers alike? *(You always get an answer. For the "and then" operation, the answer is one of the six symmetry operations. For multiplication, the answer is another number. Also, R_{360} acts like multiplying by 1: they both do nothing.)*

> In what ways are the tables for two operations different? *(There are only six symmetry operations, but there are infinitely many numbers. Also, the multiplication table is symmetric about a diagonal from the upper left to the lower right; the "and then" table is not.)*

4.2 • Transforming Squares

In this problem, students will analyze all the possible symmetries of a square and then explore the results of combinations of symmetry transformations, trying to find a single transformation that is equivalent to each combination of two transformations. They will organize their results into a table, look for patterns, and compare the results with those found for an equilateral triangle and with those for the operation of multiplication.

Launch

Explain to the class that a transformation table can be made for *any* polygon or design that has reflectional or rotational symmetry. In this problem, students will be working with a square. To introduce the problem, draw a square on the board or the overhead.

> What lines of symmetry does a square have? *(A square has four lines of symmetry: one along each diagonal, a vertical line through the center, and a horizontal line through the center.)*

> Does a square have any other kinds of symmetry? *(A square also has rotational symmetry of 90°, 180°, 270°, and 360°.)*

Distribute a paper copy of square *ABCD* (cut from the blackline master in Additional Resources), Labsheet 4.2A, and Labsheet 4.2B to each student. Have students label the back side of each vertex of their paper square.

Part A of the problem asks students to draw all the lines of symmetry on the square shown at the top of Labsheet 4.2B. You may choose to do this part as a class or to have students work in their groups. To facilitate sharing of results later, ask that everyone label the vertical line as line 1 and continue labeling the lines in a clockwise direction. Their completed drawings should look like the one below.

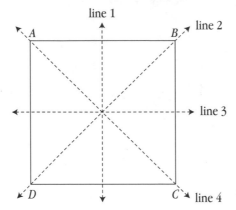

Next, have students record the results of each of the eight transformations on Labsheet 4.2A, as they did for the equilateral triangle. The completed Labsheet 4.2A should look as follows:

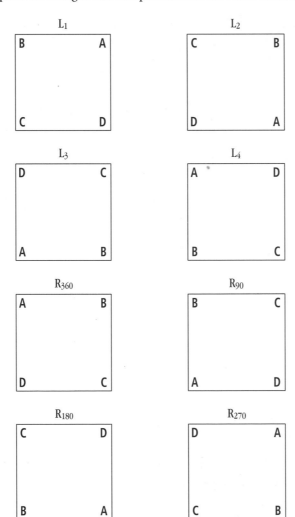

Have students work in groups of two to four to finish the problem and the follow-up. Labsheet 4.2B contains the beginnings of the tables students are to construct.

Explore

Ask that students collaborate with their groups to determine the entries, with each student filling in his or her own table. You may want to distribute transparencies of Labsheet 4.2B to one or two groups for copying the tables from the problem and follow-up for use in the class discussion.

Having students enter the row and column heads in an order similar to that used in Problem 4.1 for the triangle will facilitate discussion of similarities and differences.

Summarize

As a class, discuss the entries in the table for the symmetry transformations of a square, and resolve any differences. Then talk about the patterns in the table.

$*$	R_{360}	R_{90}	R_{180}	R_{270}	L_1	L_2	L_3	L_4
R_{360}	R_{360}	R_{90}	R_{180}	R_{270}	L_1	L_2	L_3	L_4
R_{90}	R_{90}	R_{180}	R_{270}	R_{360}	L_2	L_3	L_4	L_1
R_{180}	R_{180}	R_{270}	R_{360}	R_{90}	L_3	L_4	L_1	L_2
R_{270}	R_{270}	R_{360}	R_{90}	R_{180}	L_4	L_1	L_2	L_3
L_1	L_1	L_4	L_3	L_2	R_{360}	R_{270}	R_{180}	R_{90}
L_2	L_2	L_1	L_4	L_3	R_{90}	R_{360}	R_{270}	R_{180}
L_3	L_3	L_2	L_1	L_4	R_{180}	R_{90}	R_{360}	R_{270}
L_4	L_4	L_3	L_2	L_1	R_{270}	R_{180}	R_{90}	R_{360}

Next, discuss the similarities and differences between the tables for the symmetry transformations of an equilateral triangle and a square. Make sure the following observations are brought out:

- Each combination of transformations is equivalent to a single transformation.

- R_{360} is a "do nothing" operation; it essentially has no effect.

- R_{360} occurs once in every row and once in every column, as does every other symmetry.

- Combining two rotations or two reflections is equivalent to a rotation.

- Combining a reflection and a rotation is equivalent to a reflection.

Finally, talk about the similarities and differences between the table for symmetry transformations of a square and the multiplication table.

In this problem, students review three important properties of addition and multiplication of real numbers—the commutative property, the identity property, and the inverse property. They are then asked to analyze whether the operation they have been studying—the "and then" operation of combining symmetry transformations—satisfies some or all of these properties. The commutative property, which students encountered in the unit *Say It with Symbols,* takes on deeper meaning when students discover that not all operations are commutative.

Launch

Read or discuss the information about properties of real numbers that precedes Problem 4.3. The discussion is reproduced on Transparency 4.3A.

> In *Say It with Symbols,* you learned about the commutative property. Can anyone remind us what this property is about? *(It says that the order in which numbers are added or multiplied does not affect the result.)*

Introduce the concept of identity elements.

> When you multiply a number by 1, what do you get? *(You always get that number.)* Is there any number that you can *add* to any other number and get that number? *(Yes; you can add 0 to any number and always get that number.)*

> These numbers have special names: we call 0 and 1 *identity elements.* Zero is the *additive identity,* and 1 is the *multiplicative identity.*

The discussion of identity elements can lead into the concept of inverses.

> When I multiply 2 by $\frac{1}{2}$, what do I get? *(1)* When I multiply 213 by $\frac{1}{213}$, what do I get? *(1)*

> Remember that 1 is the multiplicative identity. Does *every* number have a number by which it can be multiplied to give a product of 1? In other words, is *every* number related to another number that you can multiply by and get back to 1, the identity? *(yes, the inverse of that number)*

> We call $\frac{1}{a}$ the *multiplicative inverse* of a because $a \times \frac{1}{a} = 1$.

> Now let's consider the operation of addition. Remember that 0 is the additive identity. What do you add to 2 to get a sum of 0? *(⁻2)* What do you add to 13.7 to get a sum of 0? *(⁻13.7)* Is every number related to another number that you can add to it and get back to 0, the identity? *(yes, the opposite of that number)*

> We call ⁻a the *additive inverse* of a because $a +$ ⁻$a = 0$.

We can talk about these properties for addition and multiplication. Now think about the operation you have been studying in this investigation, the "and then" ($*$) operation for combining symmetry transformations. Do you think these properties apply to the "and then" operation?

In this problem, you will search for the identity for the "and then" operation. You will also explore whether each element in such a table has an inverse.

Have students explore the problem and the follow-up in pairs.

Explore

A think-pair-share approach works well for this problem. Each individual reads and answers parts A, B, and C, and then partners share their responses. They continue their collaboration as they do the follow-up questions.

Summarize

Discuss the problem and the follow-up as a class. First, help students analyze the "and then" operation for an equilateral triangle to see whether it is commutative.

Give me some examples of pairs of symmetry transformations that are equivalent to the same single transformation no matter in which order they are performed.

As students offer their ideas, fill in a symmetry transformation table at the board or the overhead. If you make your table a square, it will be easier for students to look for symmetry. The following transformations fit this pattern:

- Any symmetry transformation combined with R_{360}. This accounts for the entries in the first row and the first column.

- Any symmetry transformation combined with itself. This accounts for entries along the diagonal from the upper left to the lower right.

- R_{120} and R_{240} are equivalent to the same transformation regardless of the order in which they are performed.

So far, the table looks as follows:

$*$	R_{360}	R_{120}	R_{240}	L_1	L_2	L_3
R_{360}	R_{360}	R_{120}	R_{240}	L_1	L_2	L_3
R_{120}	R_{120}	R_{240}	R_{360}			
R_{240}	R_{240}	R_{360}	R_{120}			
L_1	L_1			R_{360}		
L_2	L_2				R_{360}	
L_3	L_3					R_{360}

What about the remaining combinations? Does order matter for them? *(yes)* For each additional pair of transformations, show me how you know.

You might list the pairs at the board:

$L_1 * R_{120} = L_3$ $L_1 * R_{240} = L_2$ $L_2 * R_{120} = L_1$

$R_{120} * L_1 = L_2$ $R_{240} * L_1 = L_3$ $R_{120} * L_2 = L_3$

$L_2 * R_{240} = L_3$ $L_3 * R_{120} = L_2$ $L_3 * R_{240} = L_1$

$R_{240} * L_2 = L_1$ $R_{120} * L_3 = L_1$ $R_{240} * L_3 = L_2$

$L_3 * L_2 = R_{120}$ $L_2 * L_1 = R_{120}$ $L_3 * L_1 = R_{240}$

$L_2 * L_3 = R_{240}$ $L_1 * L_2 = R_{240}$ $L_1 * L_3 = R_{120}$

Let's add these to our table using a different color to distinguish them from the others.

As you complete the table, ask questions to help students focus on the symmetry across the diagonal.

How could you describe where $L_1 * R_{120}$ is in the table compared with $R_{120} * L_1$?

Draw a line through the diagonal to help emphasize the symmetry. One by one, focus on the fact that a pair and its reverse are located in the same spot on either side of this diagonal. The positions are symmetric about this diagonal.

$*$	R_{360}	R_{120}	R_{240}	L_1	L_2	L_3
R_{360}	R_{360}	R_{120}	R_{240}	L_1	L_2	L_3
R_{120}	R_{120}	R_{240}	R_{360}	L_2	L_3	L_1
R_{240}	R_{240}	R_{360}	R_{120}	L_3	L_1	L_2
L_1	L_1	L_3	L_2	R_{360}	R_{240}	R_{120}
L_2	L_2	L_1	L_3	R_{120}	R_{360}	R_{240}
L_3	L_3	L_2	L_1	R_{240}	R_{120}	R_{360}

When students see this diagonal relationship, ask:

If this operation were commutative, what would we see? (The parts on each side of the diagonal would be the same. All the entries would be symmetric about the diagonal.)

Are the entries in the multiplication table symmetric about the diagonal from the upper left to the lower right? *(yes)* So, we have a way to "see" commutativity in a table in which the column heads and the row heads are listed in the same order.

Could I check for commutativity by looking for symmetry about the diagonal from the lower left to the upper right? *(no)* Why not? *(Because the entries in symmetric locations come from different pairs of transformations. For example, the entry in the upper-left corner represents $R_{360} * R_{360}$, while the entry in the lower-right corner represents $L_3 * L_3$. Even if the entries are the same, it doesn't tell us about commutativity.)*

Now discuss identity elements and inverses.

Did you find the identity for this operation? *(yes, R_{360})* So, this symmetry transformation has no effect when you combine it with another transformation, in either order. How can I tell this from the table? *(The row and column for R_{360} just repeat the row and column heads.)*

Give me an example of a symmetry transformation and its inverse. *(R_{360} and R_{360}, R_{120} and R_{240}, R_{240} and R_{120}, and so on)*

How can you find an inverse for a symmetry transformation from the table? *(Look in the row for that symmetry and find R_{360}, the identity. Then find the column head for that entry. For example, to find the inverse of L_1, find R_{360} in the row for L_1. The column head for this entry is L_1, so $L_1 * L_1 = R_{360}$.)*

So, how could you check a table for the "and then" operation quickly to see whether each element has an inverse? *(You could look to see if there is an R_{360} entry in each row and then look at the corresponding entry on the other side of the diagonal; it should also be R_{360}.)*

Use the follow-up to review how to check a table for commutativity, an identity, and inverses. Then ask:

Summarize how the "and then" operation on symmetry transformations is similar to and different from the multiplication operation on whole numbers. *(When you combine two transformations, you always get a transformation. When you multiply two numbers, you always get another number. Each element has an inverse, and each operation has an identity, but the "and then" operation for symmetry transformations is not commutative.)*

Additional Answers

Answers to Problem 4.1 Follow-Up

1. Possible answer: Each cell contains one of the six symmetry transformations, which means that every combination of two transformations is equivalent to a single transformation. The upper-left and lower-right sections are filled with rotations, and the lower-left and upper-right sections are filled with reflections. Specifically, combining a rotation and a reflection is equivalent to a single reflection; combining two rotations or two reflections is equivalent to a single rotation. Also, every row and every column contains each transformation once, and there is a row and a column that contain entries that are identical to the row heads and column heads.

2. Possible answer: In both tables, the entries in the first row match the column heads and the entries in the first column match the row heads. Every entry in the table for the "and then" operation is one of the six transformations that appear in the row and column heads; the entries in the multiplication table consist of more numbers than just those in the row and column heads. The multiplication table has symmetry with respect to the diagonal from the upper left to the lower right; that is not true for the "and then" table. (Note: As will be discussed in Problem 4.3, this indicates that the multiplication operation is commutative.)

×	1	2	3	4	5	6
1	1	2	3	4	5	6
2	2	4	6	8	10	12
3	3	6	9	12	15	18
4	4	8	12	16	20	24
5	5	10	15	20	25	30
6	6	12	18	24	30	36

Answers to Problem 4.2

A. A square has four lines of symmetry and can be rotated 90°, 180°, 270°, or 360° to match the original figure.

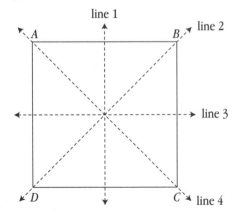

Answers to Problem 4.2 Follow-Up

2. Possible answer: In both tables, the entries in the first row match the column heads and the entries in the first column match the row heads. Every entry in the table for the "and then" operation is one of the eight transformations that appear in the row and column heads; the entries in the multiplication table consist of more numbers than just those in the row and column heads. The multiplication table has symmetry with respect to the diagonal from upper left to lower right; that is not true for the "and then" table. (Note: As will be discussed in Problem 4.3, this indicates that the multiplication operation is commutative.)

×	1	2	3	4	5	6	7	8
1	1	2	3	4	5	6	7	8
2	2	4	6	8	10	12	14	16
3	3	6	9	12	15	18	21	24
4	4	8	12	16	20	24	28	32
5	5	10	15	20	25	30	35	40
6	6	12	18	24	30	36	42	48
7	7	14	21	28	35	42	49	56
8	8	16	24	32	40	48	56	64

Answers to Problem 4.3 Follow-Up

1. The operation $*$ is not commutative for a square; order does matter in many cases. For example, reflecting the square over line 1 and then over line 2 is equivalent to rotating it 270°; reflecting it over line 2 and then over line 1 is equivalent to rotating it 90°.

2. The identity element is R_{360}.

3. R_{360}, R_{180}, and each line reflection are inverses of themselves; R_{90} and R_{270} are inverses of each other since $R_{90} * R_{270} = R_{270} * R_{90} = R_{360}$.

ACE Answers

Extensions

11a. A regular pentagon has five lines of symmetry and can be rotated 72°, 144°, 216°, 288°, and 360° about its centerpoint to coincide with the original.

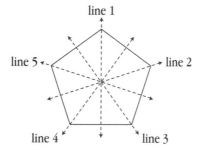

11b. (Note: If students label the lines in a different order, their tables will vary. In terms of algebraic structure, all possible tables are the same.)

$*$	R_{360}	R_{72}	R_{144}	R_{216}	R_{288}	L_1	L_2	L_3	L_4	L_5
R_{360}	R_{360}	R_{72}	R_{144}	R_{216}	R_{288}	L_1	L_2	L_3	L_4	L_5
R_{72}	R_{72}	R_{144}	R_{216}	R_{288}	R_{360}	L_4	L_5	L_1	L_2	L_3
R_{144}	R_{144}	R_{216}	R_{288}	R_{360}	R_{72}	L_2	L_3	L_4	L_5	L_1
R_{216}	R_{216}	R_{288}	R_{360}	R_{72}	R_{144}	L_5	L_1	L_2	L_3	L_4
R_{288}	R_{288}	R_{360}	R_{72}	R_{144}	R_{216}	L_3	L_4	L_5	L_1	L_2
L_1	L_1	L_3	L_5	L_2	L_4	R_{360}	R_{216}	R_{72}	R_{288}	R_{144}
L_2	L_2	L_4	L_1	L_3	L_5	R_{144}	R_{360}	R_{216}	R_{72}	R_{288}
L_3	L_3	L_5	L_2	L_4	L_1	R_{288}	R_{144}	R_{360}	R_{216}	R_{72}
L_4	L_4	L_1	L_3	L_5	L_2	R_{72}	R_{288}	R_{144}	R_{360}	R_{216}
L_5	L_5	L_2	L_4	L_1	L_3	R_{216}	R_{72}	R_{288}	R_{144}	R_{360}

11c. The $*$ operation is not commutative for this set of transformations. For example, $L_1 * L_2 = R_{216}$ while $L_2 * L_1 = R_{144}$.

11d. R_{360}

11e. R_{360} and R_{360}; R_{72} and R_{288}; R_{144} and R_{216}; L_1 and L_1; L_2 and L_2; L_3 and L_3; L_4 and L_4; L_5 and L_5

Part 1: Creating Tessellations

Materials: grid paper, isometric dot paper, stiff paper, an angle ruler or protractor, transparent tape, and colored pencils or markers

Directions:

- Draw square *PQRS* on grid paper.
- Draw triangle 1 as shown.
- Draw triangle 1′, the image of triangle 1 under a 270° rotation about point *P*.
- Draw triangle 2 as shown.
- Draw triangle 2′, the image of triangle 2 under a 270° rotation about point *R*.

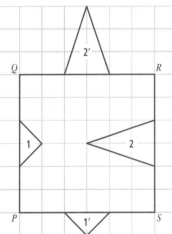

Now you will start with a copy of square *PQRS* and perform the same rotations as before. However, this time you will cut out triangles 1 and 2 and rotate the cut pieces.

- Copy square *PQRS* and triangles 1 and 2 onto a stiff sheet of paper. Cut out square *PQRS*.
- Cut out triangle 1, rotate it 270° about point *P*, and tape it in place.
- Cut out triangle 2, rotate it 270° about point *R*, and tape it in place.

Trace copies of your finished shape to create a tessellation. Recall that a *tessellation* is a design made by fitting together copies of a basic shape without gaps or overlaps. You can decorate the basic shapes to make your tessellation more interesting.

Assigning the Unit Project

The optional unit project consists of two hands-on activities. Either or both parts of the project will give students an opportunity to apply what they have learned about symmetry. The project can be started while Investigation 4 is in progress.

In part 1, students create tessellating shapes by applying symmetry transformations to the sides of a square and a nonsquare rhombus. In part 2, students explore the symmetries of the various shapes appearing in an origami construction.

Each project can be done by students individually, but students will make more discoveries about tessellations and symmetry if they share ideas about their work in groups or as a whole class.

A guide to the project can be found in the Assessment Resources section.

We say that the shape below *tessellates* because it can be used to create a tessellation.

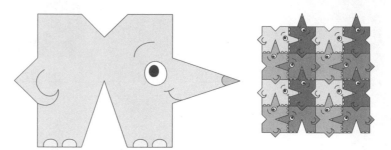

Starting with a square, try to create your own tessellating shape. Cut out pieces from the square, rotate them, and tape them in place. As you work, think about this question: *Will any cutout shape and any rotation turn a square into a shape that tessellates?* When you find a shape that tessellates, trace it several times to make a tessellation. Tape your basic shape to the paper.

Now see if you can create a tessellating shape from a rhombus with angles of 60° and 120°. You may find it easier to make your design on isometric dot paper and then cut out a copy from stiff paper. As you work, think about this question: *Will any cutout shape and any rotation turn a rhombus with 60° and 120° angles into a shape that tessellates?* If you find a shape that tessellates, trace it several times to make a tessellation. Tape your basic shape to the paper. If you were unable to create a shape that tessellates, explain why you think your efforts were unsuccessful.

Part 2: Making a Wreath and a Pinwheel

Origami is the Japanese art of paper folding. In this part of the project, you will make an origami wreath and then transform your wreath into a pinwheel.

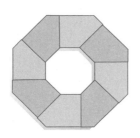 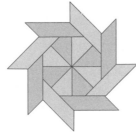

Materials: eight paper squares of the same size (four paper squares in each of two colors gives a nice result)

Directions:

The wreath is made by connecting eight folded squares. Follow these instructions to fold each square:

- Fold a paper square to create the creases shown.

- Fold down the top corners of the square to make a "house," and then fold the house in half so that the flaps are on the inside.

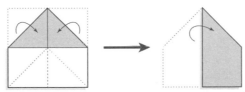

- Hold the "half-house" at its point, and push the bottom corner in along the folds to make a parallelogram.

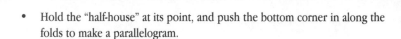

Hold here.   Push here.

Follow these steps to connect the eight pieces:

- Position two of the folded pieces as shown on the left below. (If you used different colors, use one piece of each color.) Slide the point of the right piece into the fold pocket of the left piece.

folded edge   folded edge

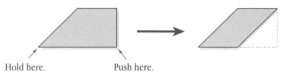

- Fold down the tips that extend over the inserted piece, and tuck them into the valley formed by the folds of the inserted piece.

Fold these tips inward.   Tuck the tips into this pocket.

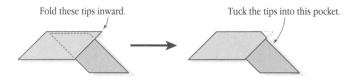

- Follow the steps above to attach the remaining folded pieces.
- Complete the wreath by connecting the last piece to the first piece, being careful to fold each flap over only one folded layer.

- To make a pinwheel, gently slide the sides of the wreath toward the center.

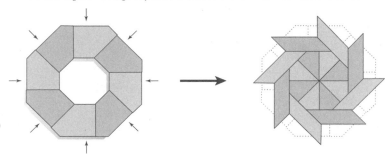

Study the drawings in the instructions for creating the origami wreath and pinwheel, and look for symmetries in the figures created at each stage.

1. Describe the reflectional and rotational symmetries of
   a. the square.
   b. the "house."
   c. the "half-house."
   d. the parallelogram.
   e. the completed wreath.
   f. the pinwheel.

2. Slide your pinwheel back into a wreath shape. If you gently push on a pair of opposite sides of your wreath, you will get a pinwheel with only two "wings." Describe the reflectional and rotational symmetries of this figure.

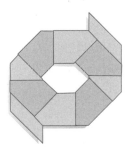

3. Slide your pinwheel back into a wreath shape. Gently push on opposite sides of the pinwheel to produce other shapes. Look at the shape of the center opening. What shapes can you create by pushing on the sides of the wreath?

Assessment
Resources

Students will need access to tracing paper, reflecting devices, rulers, and angle rulers or protractors for all the assessment pieces. For the unit test, each student will need a copy of the regular hexagon below. A blackline master with several copies of this hexagon can be found on page 178.

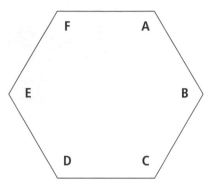

Check-Up 1

1. **a.** Does the word below have reflectional symmetry? If so, draw all the lines of symmetry.

WOW

   b. Change one letter in the word to form a word with rotational symmetry.

   c. How could you check that your word in part b has rotational symmetry?

   d. Where is the center of rotation for your word? What is the angle of rotation?

In 2–4, match the design with the correct description from parts a–c. Then draw the lines of symmetry, and indicate the centers and angles of rotation for the designs.

2. _____ 3. _____ 4. _____

   a. This design has rotational symmetry, but no reflectional symmetry.

   b. This design has reflectional symmetry, but no rotational symmetry.

   c. This design has both reflectional and rotational symmetry.

Quiz

1. a. Outline a basic design element that could be copied and transformed to create the tessellation at right.

    b. Describe the symmetries of your basic design element.

    c. Use what you know about symmetry transformations—reflections, rotations, and translations—to describe how the entire design could be created from your basic design element.

2. a. Outline a basic design element that could be copied and transformed to create the tessellation below.

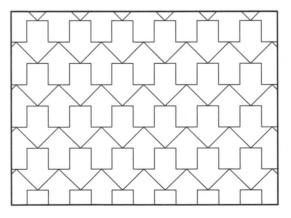

Quiz

b. Describe the symmetries of your basic design element.

c. Use what you know about symmetry transformations—reflections, rotations, and translations—to describe how the entire design could be created from your basic design element.

3. Describe a transformation or a series of transformations you could apply to this spaceship so that the final image is completely covered by the cloud of gases. Give the reflection line for each reflection, the center and angle of rotation for each rotation, and the length and direction for each translation.

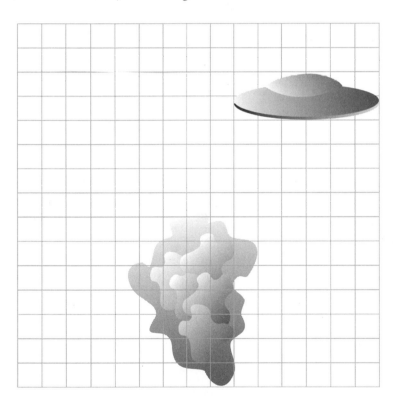

Quiz

4. a. Use a ruler and an angle ruler or protractor to draw the image of square *ABCD* under a reflection over the line.

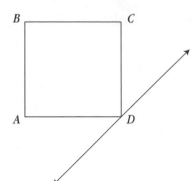

b. Describe the relationship between a point on square *ABCD* and the image of that point under the reflection.

5. a. Use a ruler and an angle ruler or protractor to draw the image of square *ABCD* under a 45° rotation about point *A*.

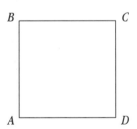

b. Describe the relationship between a point on square *ABCD* and the image of that point under the rotation.

6. a. Reflect triangle *ABC* over the line, and then rotate the image 180° about point *D*.

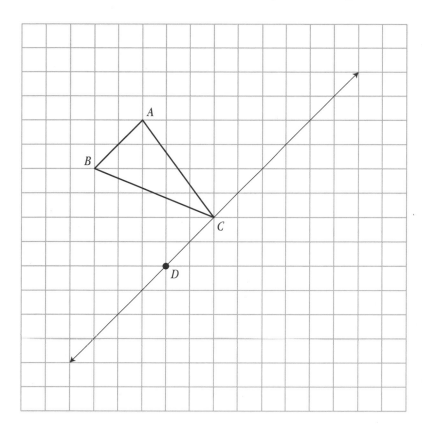

b. Precisely describe a single transformation that would carry triangle *ABC* onto the final image from part a.

Name _____ Date _____

1. a. Under a particular transformation, A' is the image of A and B' is the image of B. Give a detailed description of the transformation.

    b. Use a ruler and an angle ruler or protractor to help you draw the image of trapezoid $ABCD$ under the transformation you described in part a.

2. a. On the grid at right, draw the shape that these commands would create:

        ```
        Draw:
        Line [(-1, 3), (4, 7)]
        Line [(4, 7), (8, 7)]
        Line [(8, 7), (8, 1)]
        Line [(8, 1), (1, 3)]
        Line [(1, 3), (-1,3)]
        ```

    b. Using another color or dashed lines, draw the shape that these commands would create:

        ```
        Draw:
        Line [(3, -1), (7, 4)]
        Line [(7, 4), (7, 8)]
        Line [(7, 8), (1, 8)]
        Line [(1, 8), (3, 1)]
        Line [(3, 1),(3, -1)]
        ```

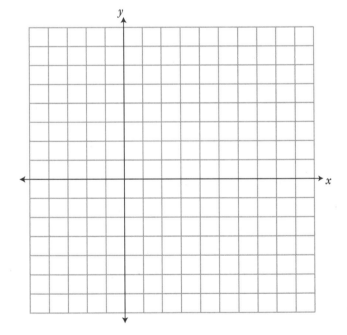

    c. For what transformation is the shape from part b the image of the shape from part a?

Check-Up 2

d. The same transformation is applied to a different shape, and the result is the shape created by these commands:

```
Draw:
Line [(-1, -3), (3, 1)]
Line [(3, 1), (3, 5)]
Line [(3, 5), (-1, -3)]
```

On the grid below, draw the *original* shape, and explain how you made your drawing.

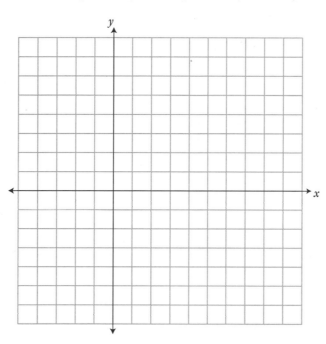

Assign these questions as additional homework, or use them as review, quiz, or test questions.

1. Describe how you could move the word ME from its original position to the final position in four reflections. Draw the result of each reflection.

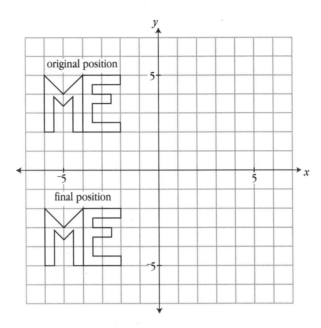

2. The shape below can be moved from its original position to its final position in one rotation. Describe the rotation.

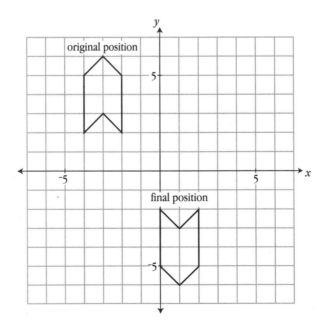

3. a. Use an angle ruler or a protractor to help you draw the image of this figure under a 90° rotation about point *K*.

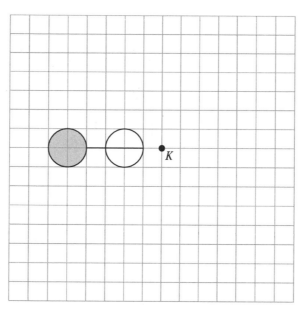

b. Use an angle ruler or a protractor to help you draw the image of this figure under a 120° rotation about point *P*.

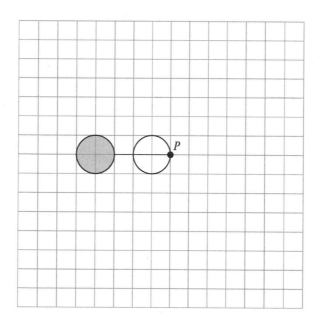

4. Sketch the final result of rotating the arrow 120° about point *P* and then reflecting the image over the line of reflection.

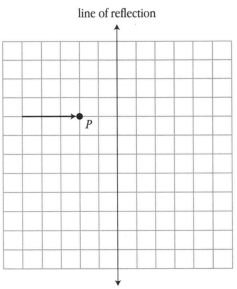

5. **a.** On the grid below left, perform a reflection on the shape and a second reflection on its image so that the combination of reflections is equivalent to a translation. Describe the two reflections and the equivalent translation.

   b. On the grid below right, perform a reflection on the shape and a second reflection on its image so that the combination of reflections is equivalent to a rotation. Describe the two reflections and the equivalent rotation.

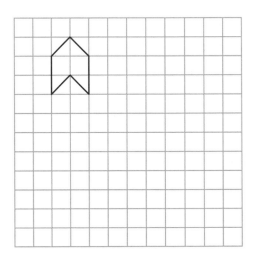
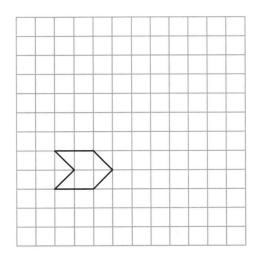

6. a. On a blank sheet of paper, create a tessellation from copies of this basic design element. Number the copies to indicate the order in which you drew them.

b. Explain how you slid the basic design element from one position to another to make your tessellation. For example, in what direction and for what distance did you slide the shape to make the first copy? For how many copies did you continue in this direction? After how many copies did you change direction?

7. a. This "snowflake" was constructed by folding a hexagonal piece of paper and making cuts. The gray sections represent areas that have been cut out. Describe, with words and illustrations, the steps required to fold and cut a hexagonal sheet of paper to make this design.

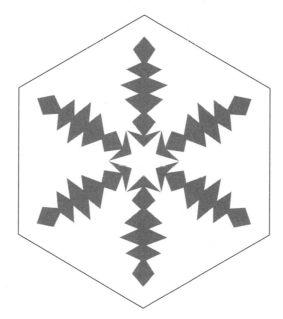

b. Use your description from part a to make the snowflake.

Name _____ Date _____

1. A graphic artist tried to translate a copy of the original school bus drawing below, but he accidentally left one of the windows behind.

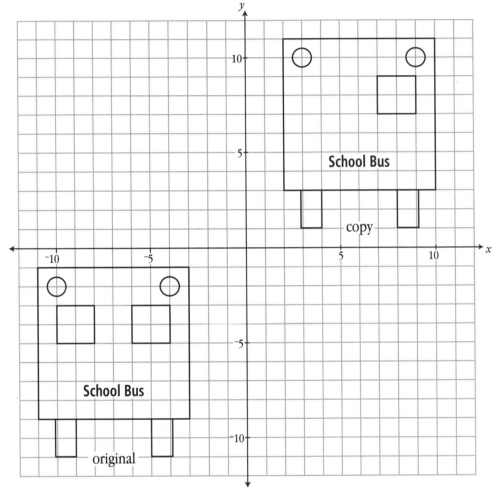

   a. Make a table showing the coordinates of the vertices of the left window of the original bus and the coordinates of the vertices that this window should have in the copy.

   b. Describe the translation so that someone else could start with the drawing of the original bus and draw the correct image.

2. **a.** The large triangle below is made from identical smaller triangles. Describe the symmetries of one of the small triangles.

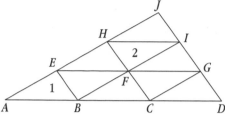

b. If you moved triangle 1 so it fit exactly on triangle 2, which vertices would be matched?

c. Carefully describe a combination of transformations that would move triangle 1 so that it exactly matches triangle 2. You may add lines or points to the diagram if you need to.

d. Describe a single transformation that will match triangle 1 with triangle 2.

3. **a.** Describe all the symmetries of triangle 1.

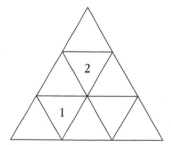

b. Because an equilateral triangle has so many symmetries, there are many ways to move triangle 1 so that it exactly matches triangle 2. Describe three ways to move triangle 1 to match triangle 2.

4. A regular hexagon has six lines of symmetry, and it can be rotated 60°, 120°, 180°, 240°, 300°, and 360° to coincide with the original.

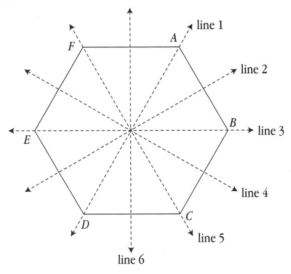

Your teacher has given you a copy of the hexagon above. Copy the vertex labels onto the back of the hexagon. Use your paper hexagon to help you complete the table below to show the results of combining pairs of symmetry transformations. Remember, each entry should be the result of performing the transformation in the left column followed by the transformation in the top row.

∗	R_{60}	R_{120}	L_1	L_2	L_3	L_4	L_5	L_6
R_{60}		R_{180}	L_2			L_5		L_1
L_1	L_6	L_5		R_{300}			R_{120}	

Unit Test

5. Points A' and B' are the images of points A and B under a particular transformation.

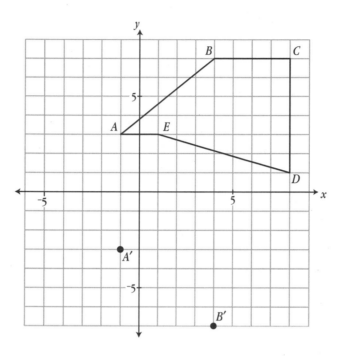

a. Give the coordinates of points A, B, A', and B'.

b. Describe the transformation in words.

c. Write a general rule for transforming a point (x, y) under the transformation.

d. Give the coordinates of the images of points C, D, and E under the transformation, and draw the image of polygon $ABCDE$ on the grid above.

Notebook Checklist

Journal Organization

_____ Problems and Mathematical Reflections are labeled and dated.

_____ Work is neat and is easy to find and follow.

Vocabulary

_____ All words are listed.   _____ All words are defined or described.

Check-Ups and Quiz

_____ Check-Up 1   _____ Check-Up 2

_____ Quiz

Homework Assignments

_____ _____

_____ _____

_____ _____

_____ _____

_____ _____

_____ _____

_____ _____

_____ _____

_____ _____

_____ _____

_____ _____

_____ _____

_____ _____

_____ _____

© Dale Seymour Publications®

Self-Assessment

Vocabulary

Of the vocabulary words I defined or described in my journal, the word _____ best demonstrates my ability to give a clear definition or description.

Of the vocabulary words I defined or described in my journal, the word _____ best demonstrates my ability to use an example to help explain or describe an idea.

Mathematical Ideas

In *Kaleidoscopes, Hubcaps, and Mirrors,* I learned about symmetry and transformations.

1. **a.** I have learned the following about . . .

   . . . symmetry:

   . . . reflections, rotations, and translations:

   . . . combining transformations:

   b. Here are page numbers of journal entries that give evidence of what I have learned, along with descriptions of what each entry shows:

2. **a.** These are the mathematical ideas I am still struggling with:

   b. This is why I think these ideas are difficult for me:

   c. Here are page numbers of journal entries that give evidence of what I am struggling with, along with descriptions of what each entry shows:

Class Participation

I contributed to the class discussion and understanding of *Kaleidoscopes, Hubcaps, and Mirrors* when I . . . (Give examples.)

Answers to Check-Up 1

1. **a.** The word has reflectional symmetry over the line shown below.

b.

c. Possible answer: You could trace the word and then rotate the tracing.

d. The center of rotation is the center of the "O." The angle of rotation is 180°.

2. b

3. c

4. a

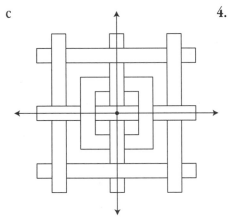

180° angle of rotation

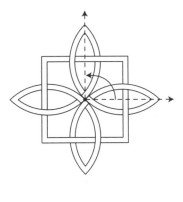

90° angle of rotation

Answers to the Quiz

1. **a.** Possible answer:

b. The design element shown in part a has rotational symmetry about the centerpoint with a 120° angle of rotation.

c. Possible answer: The basic design element can be rotated 60° as shown here:

The image can then be rotated 60° about the same point, and the process can be continued to create a flower shape made of six basic design elements. The flower shape can then be translated, according to the arrow shown, to create a diagonal strip.

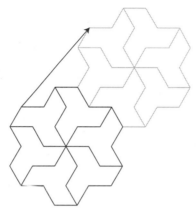

Then the entire strip can be translated, according to the arrow shown, to create the entire pattern.

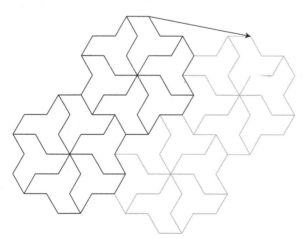

2. a. Possible answer:

b. The arrow shape has reflectional symmetry over a vertical line through its point.

c. Possible answer: The basic design element can be translated to create a row of arrows. Then, as shown in the illustration below, this row can be reflected over the line through the bottoms of the arrows and then translated as indicated by the arrow.

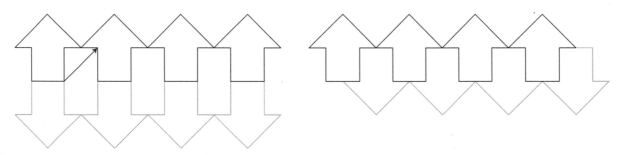

3. Possible answer: Rotate the ship 90° about point *P*, and then translate the image 10.5 units down and 1.5 units to the left. (Note: There are many other correct transformations that students might suggest.)

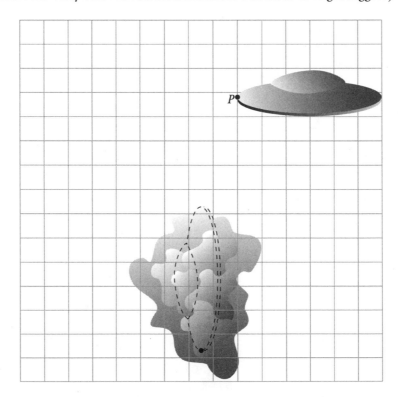

4. a.

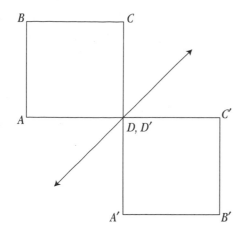

b. The image of a point on square *ABCD* lies on the line that is perpendicular to the line of reflection and passes through the point. The point and its image are on opposite sides of the line of reflection, and they are the same distance from the line of reflection.

5. a.

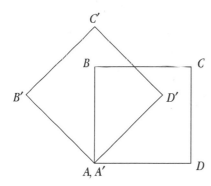

b. A point and its image are the same distance from point *A*. The measure of the angle formed by the point, point *A*, and the image point is 45°.

6. a. The final image is labeled A″B″C″ in the drawing below.

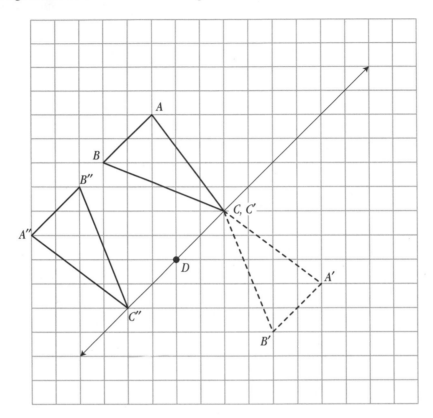

b. a reflection over a line perpendicular to the line of reflection and passing through point D

Answers to Check-Up 2

1. a. The transformation is a translation of approximately 3.9 cm at an angle of about 20° from the horizontal. Students may have other ways to describe the direction.

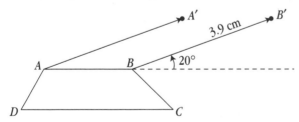

b.

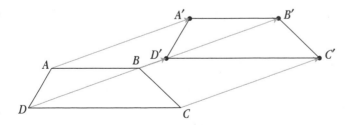

2. **a.** These commands would create the shape drawn with solid lines on the grid below.

    b. These commands would create the shape drawn with dashed lines on the grid below.

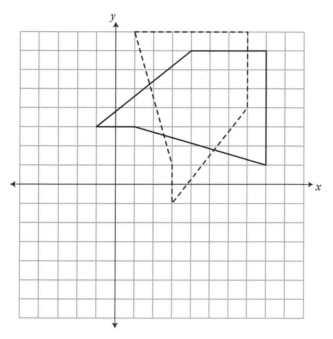

    c. a reflection over the line $y = x$

    d. The original shape is drawn with solid lines on the grid below. Students may use the coordinate rule $(x, y) \rightarrow (y, x)$ to find the coordinates of the original shape, or they may follow the commands to draw the image and then use a reflecting device or tracing paper to find the reflection.

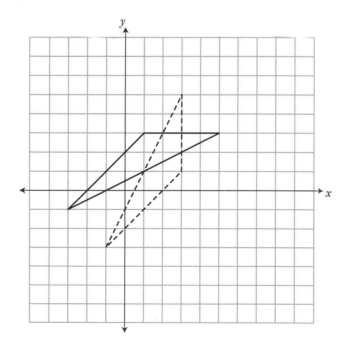

Answer Keys

Answers to the Question Bank

1. Possible answer: Reflect the original over the *y*-axis, reflect the image over the *x*-axis, reflect the second image over the *y*-axis, and reflect the third image over the line shown.

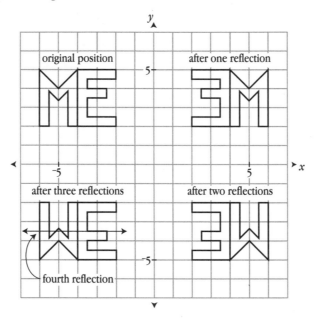

2. Rotate the original shape 180° about point *X*.

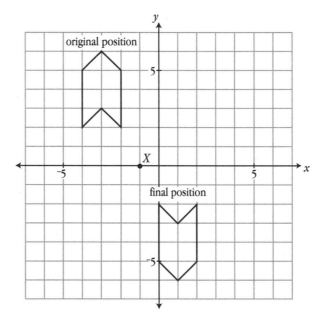

3. a.

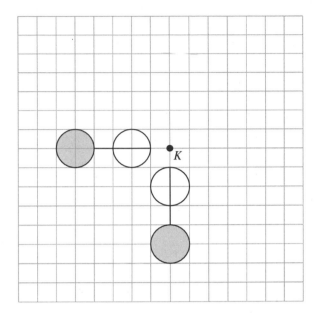

b.

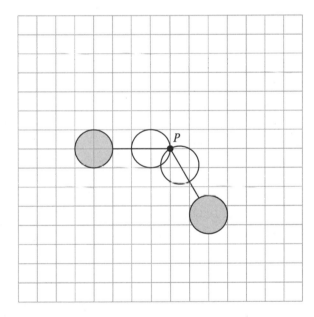

4.

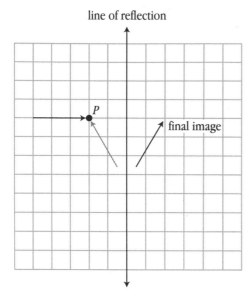

line of reflection

P

final image

5. **a.** Possible answer: In the drawing on the left, the original shape was reflected over line 1, and then the image was reflected over line 2. The combination of these reflections is equivalent to a translation of 8 units to the right.

b. Possible answer: In the drawing on the right, the original shape was reflected over line 1, and then the image was reflected over line 2. The combination of these reflections is equivalent to a rotation of 180° about point *P*.

line 1   line 2

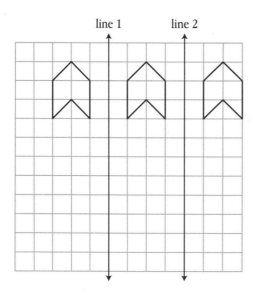

line 2

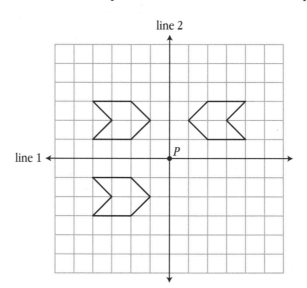

line 1

P

6. a. Tessellations should resemble the one below. The numbering of the copies will vary.

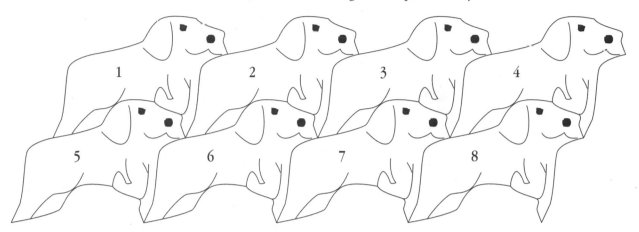

b. Answers will vary. The description should match the order in which the student numbered the copies. For the tessellation above, the shape was placed in the upper-left corner and traced to produce copy 1. The shape was then slid to the right in intervals of one dog length to make copies 2, 3, and 4. Then the shape was slid down and to the left at an angle of about 25° from the vertical to make copy 5. The shape was then slid to the left by intervals of one dog length to create copies 6, 7, and 8.

7. a. Possible answer: Fold the hexagon in half, and then fold the half into thirds.

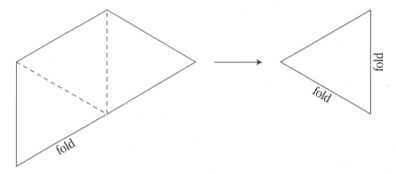

Cut out the same pattern from each of the two folded sides.

Alternatively, students might fold the hexagon in half, then in thirds, and then in half again to make a shape with 12 layers. In this case, it is only necessary to cut along one side of the triangle.

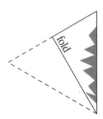

Answers to the Unit Test

1. a.

Coordinates of original window	Coordinates of copy
$(^-10, ^-3)$	$(3, 9)$
$(^-10, ^-5)$	$(3, 7)$
$(^-8, ^-3)$	$(5, 9)$
$(^-8, ^-5)$	$(5, 7)$

b. Possible answer: The translation matches any point (x, y) on the original bus to a point $(x + 13, y + 12)$ on the copy.

2. a. The small triangle has no symmetries.

b. *A* and *I*; *B* and *H*; *E* and *F*

c. Possible answers: Translate triangle 1 horizontally until point *E* matches point *F* and then rotate it 180° about point *F*. Or, rotate triangle 1 180° about point *E* and then translate it horizontally until it matches triangle 2. Or, reflect triangle 1 over line *EG* and then reflect it over the vertical line shown below.

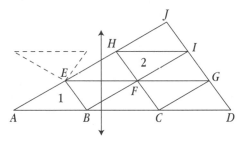

d. a rotation of 180° about the midpoint of segment *EF*

3. a. The equilateral triangle has rotational symmetry with an angle of rotation of 60°. It has reflectional symmetry with three lines of symmetry, each passing through a vertex and the midpoint of the opposite side.

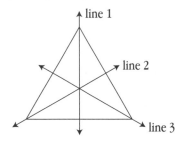

b. Possible answer: You could rotate triangle 1 240° about the vertex that touches triangle 2. You could reflect triangle 1 over its horizontal side and then reflect the image over the left side of triangle 2. You could translate triangle 1 until its upper-right vertex matches the upper-right vertex of triangle 2.

4.

✳	R_{60}	R_{120}	L_1	L_2	L_3	L_4	L_5	L_6
R_{60}	R_{120}	R_{180}	L_2	L_3	L_4	L_5	L_6	L_1
L_1	L_6	L_5	R_{360}	R_{300}	R_{240}	R_{180}	R_{120}	R_{60}

5. a. $A(-1, 3); B(4, 7); A'(-1, -3); B'(4, -7)$

  b. a reflection over the x-axis

  c. $(x, y) \rightarrow (x, -y)$

  d. $C'(8, -7); D'(8, -1); E'(1, -3)$

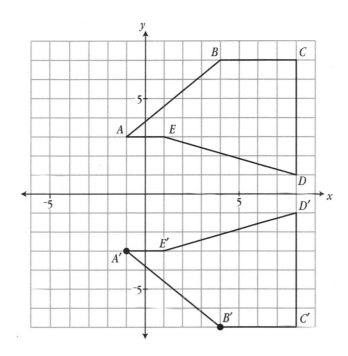

The first assessment piece for *Kaleidoscopes, Hubcaps, and Mirrors* is the check-up. A suggested scoring rubric and grading scale for the check-up are presented here.

Suggested Scoring Rubric

This rubric employs a scale with a total of 10 possible points. You may use the rubric as presented here or modify it to fit your district's requirements for evaluating and reporting students' work and understanding.

question 1: 4 points

- *part a:* 1 point for identifying the line of symmetry (No credit is given for just stating that the drawing has reflectional symmetry, as this could be a guess. Understanding is demonstrated through a drawing showing a correct line of symmetry.)
- *part b:* 1 point for identifying the correct letter to be changed and the letter to change it to
- *part c:* 1 point for describing a method (tools and/or process) for rotating the drawing so that it could be checked for symmetry
- *part d:* 1 point for identifying both the center of rotation and the angle of rotation (Only 1 point is given here because the two ideas are connected; both are needed to show understanding of the concept.)

questions 2, 3, and 4: 2 points each

- 1 point for a correct match
- 1 point for correctly drawing lines of symmetry and identifying centers and angles of rotation

Grading Scale

Points	Grade
9 to 10	A
8	B
6 to 7	C
5	D

The second assessment piece for *Kaleidoscopes, Hubcaps, and Mirrors* is the partner quiz. A suggested scoring rubric and grading scale for the quiz are presented here.

Suggested Scoring Rubric

This rubric employs a scale with a total of 23 possible points. You may use the rubric as presented here or modify it to fit your district's requirements for evaluating and reporting students' work and understanding.

questions 1 and 2: 5 points each
- *part a:* 1 point for outlining a basic design element
- *part b:* 2 points for completely describing all symmetries (e.g. giving the angle of rotation for rotational symmetry)
- *part c:* 2 points for giving a description, in the language of symmetry transformations, that will create the entire design (1 point for a partial description)

question 3: 3 points
- 3 points for giving transformations that would produce a final image that would be completely covered by the cloud (2 points for transformations that would result in a partially covered final image; 1 point for transformations that would produce a final image near the cloud)

question 4: 3 points
- *part a:* 1 point for drawing a correct image
- *part b:* 2 points for describing that an image point lies on a line perpendicular to the line of reflection through the point and that the point and its image are the same distance from the line of reflection (1 point for a partial description)

question 5: 3 points
- *part a:* 1 point for drawing a correct image
- *part b:* 2 points for describing that a point and its image are an equal distance from point *A* and that the measure of the angle formed by the three points is 45° (1 point for a partial description)

question 6: 4 points
- *part a:* 1 point for drawing a correct image
- *part b:* 2 points for stating that a reflection is needed and that it occurs over a line perpendicular to the line of reflection and passing through point *D* (1 point for a partial description)

Grading Scale

Points	Grade
21 to 23	A
18 to 20	B
15 to 17	C
12 to 14	D

The optional unit project has two parts. Either or both can be used to give students an opportunity to apply their understanding of symmetry in a hands-on activity. In part 1, Creating Tessellations, symmetry transformations are used to guide the construction of tessellating designs. In part 2, Making a Wreath and a Pinwheel, students explore the symmetries of the various shapes appearing in an origami construction.

Part 1: Creating Tessellations

In this part of the project, students apply symmetry transformations to attempt to create tessellating shapes from a square and a nonsquare rhombus. The activity is open-ended, and students are given questions to think about as they work. After they have had some exploration time, the class can be brought together to share their discoveries.

The project begins by guiding students as they create a tessellating shape from a square. The exercise will start students thinking about tessellations and about the types of symmetry transformations that can be applied to a square to create a tessellating shape. You may want to check students' work at this stage to make sure they are on the right track.

Making a Tessellating Shape from a Square
Following the introductory activity, students make their own tessellating shape from a square by performing rotations of pieces cut from a paper square. They can then share their ideas in a class discussion.

One discovery that students might make is that they can cut any shape from one side of a square, rotate it 270° about a vertex, and attach it to the new side to create a figure that tessellates.

In addition, rotating a cutout shape 180° about the midpoint of a side will create a shape that will tessellate. The tessellating figure below was made by cutting a shape from each of three sides of the original square and rotating each cutout 180° about the midpoint of the appropriate side.

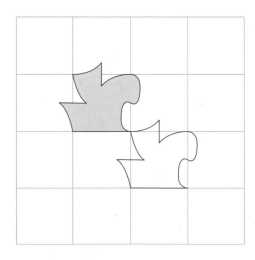

You might explore with the class the relationship between the ways that symmetry transformations can be applied to a square to create a tiling of squares, and the kinds of modifications that can be made to a square to create a new tessellating shape. Take the discussion as far as students are interested.

Making a Tessellating Shape from a Rhombus
Students next create a tessellating shape from a rhombus with angles of 60° and 120°. They will discover that they can cut any shape from one side of a rhombus with these angles, rotate it about a vertex to an adjacent side, and attach it to the new side to create a figure that tessellates.

Challenge the class to try other kinds of modifications. For example, they might try rotating a cutout shape 180° about the midpoint of a side, or rotating a cutout shape that has reflectional symmetry over a perpendicular line through the midpoint of a side. If time permits, they could also try modifying rhombuses that have other interior angles. For rhombuses with other angles, there are restrictions on the cutout shape.

Remind students that if a particular modification results in a tiling that has gaps, the tiling is by definition not a tessellation. For example, the modifications made to the rhombus below result in a shape that does not tessellate.

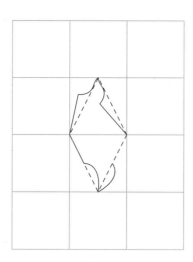
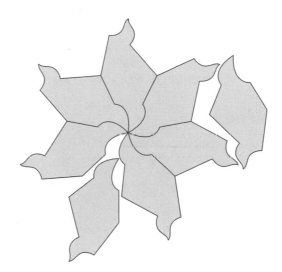

The books *Introduction to Tessellations* and *Teaching Tessellating Art* offer a wealth of ideas about exploring tessellations and symmetry transformations with your class. For more information, see the list of resources on page 11.

Part 2: Making a Wreath and a Pinwheel

Origami is an incredibly rich and complex art form. However, there are some simple but amazing figures that even a beginner can produce successfully. The wreath and pinwheel that students create in this project show again how symmetry plays a fundamental role in much design.

Students will follow the directions in their books to make a wreath from squares of paper and then transform that wreath into a pinwheel. Figures made from origami paper are easier to manipulate, but regular-weight paper will work.

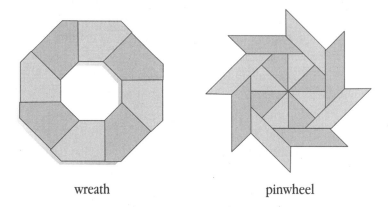

wreath   pinwheel

Remind students that the more carefully they cut out the paper squares (which must be the same size), the better their results will be. Also, sharper folds will produce better results. The tricky part of this assembly is tucking one parallelogram into another and folding the tips into the valley. The tips must not "trap" the center part of the parallelogram into which they are being folded.

Answers to the Questions

1. **a.** The square has four lines of symmetry (as indicated by the dashed fold lines) and rotational symmetries of 90°, 180°, and 270°.

    b. The "house" has one line of symmetry, a vertical line through the tip of the "roof," and no rotational symmetry.

    c. The "half-house" has no reflectional symmetry and no rotational symmetry.

    d. The parallelogram has a rotational symmetry of 180° and no reflectional symmetry.

    e. Answers will vary. For a wreath created from two alternating colors, students may say it has rotational symmetries of 90°, 180°, and 270° and no reflectional symmetry. A single-color wreath has rotational symmetries of 45°, 90°, 135°, 180°, 225°, and 270°. The octagonal figure considered as a whole does have eight lines of symmetry, but when the lines separating the individual pieces are taken into consideration, the figure does not have reflectional symmetry. Acknowledge correct reasoning, with the knowledge that some students might choose to ignore the outlines of individual pieces in their search for symmetry.

f. The pinwheel has no line symmetry. For a pinwheel created from two alternating colors, students may say it has rotational symmetries of 90°, 180°, and 270°. A single-color pinwheel has seven rotational symmetries: 45°, 90°, 135°, 180°, 225°, and 270°.

2. The two-winged pinwheel has no reflectional symmetry and one rotational symmetry of 180°.

3. Answers will vary. The center can take many shapes, including regular and non-regular quadrilaterals, hexagons, and octagons.

Blackline Masters

Problem 1.1

A.

B.

C.

D.

Problem 1.2

hubcap 1

hubcap 2

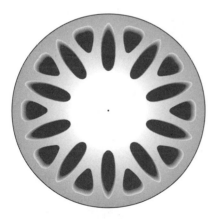

hubcap 3

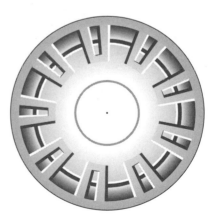

hubcap 4

Problem 1.3

design 1

design 2

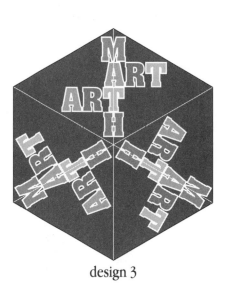

design 3

design 4

design 5

design 6

Problem 1.4

tessellation 1

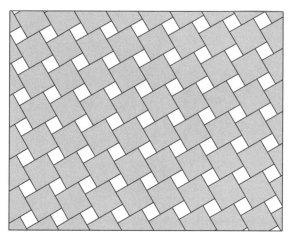

tessellation 2

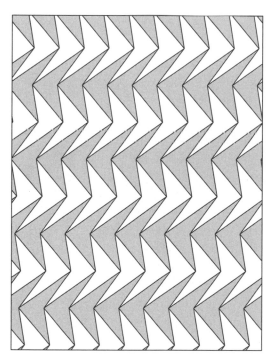

tessellation 3

tessellation 4

ACE Questions 1–3

1.

2.

3.

ACE Questions 4–8

4.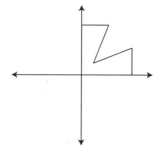

5. COOKIE

6.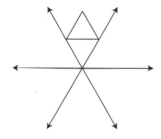

7.   1 2 3 4 5 6 7 8 9 0

8.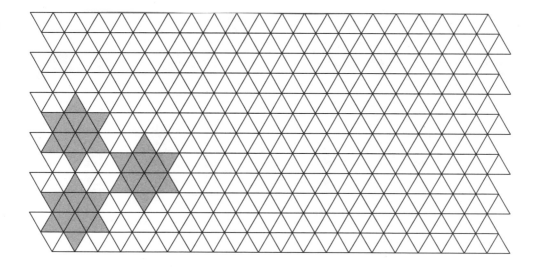

ACE Questions 9–12

9.

10.

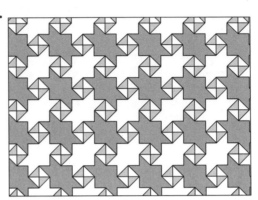

11.

12.

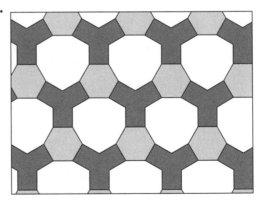

ACE Questions 20–23

20.

21.

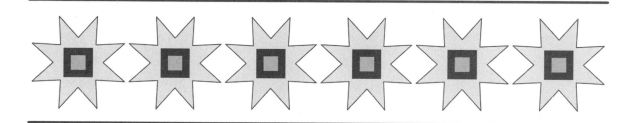

22.

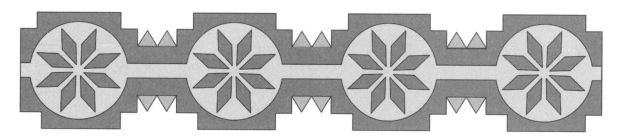

23.

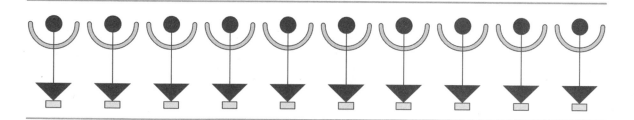

Problem 2.1

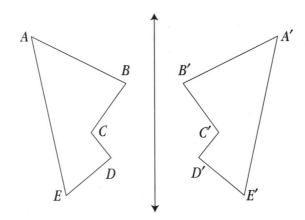

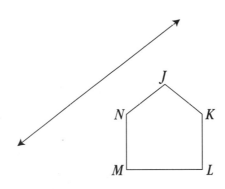

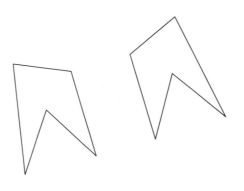

Problem 2.1 Follow-Up

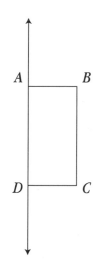

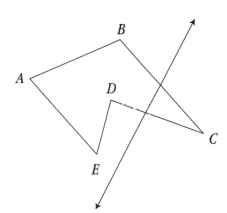

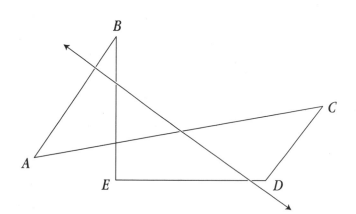

Problem 2.2 and Follow-Up

diagram 1

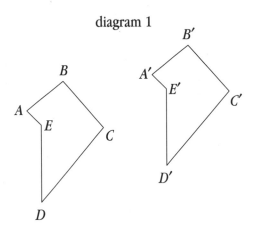

diagram 2

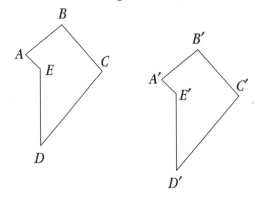

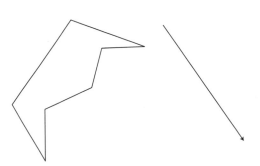

Problem 2.3

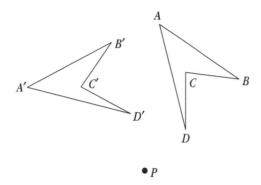

● P

Rotate 90° about point P.

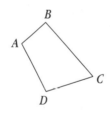

● P

Rotate 45° about point Q.

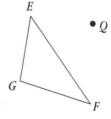

● Q

Problem 2.3 Follow-Up

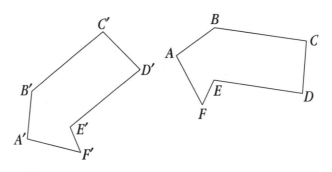

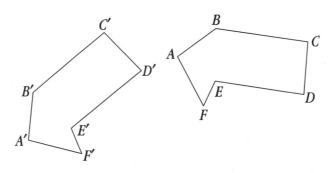

Problem 2.4

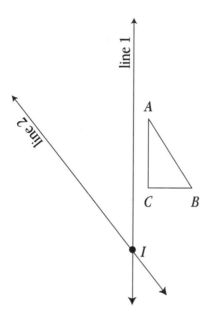

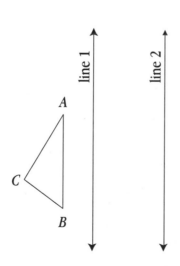

Kaleidoscopes, Hubcaps, and Mirrors

Problem 2.4 Follow-Up

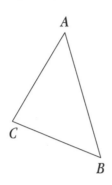

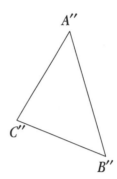

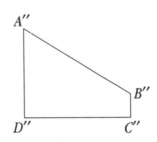

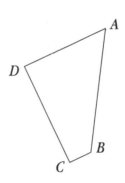

ACE Questions 1–4

1.

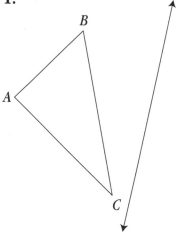

2.

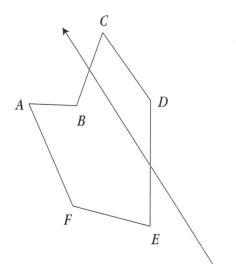

3.

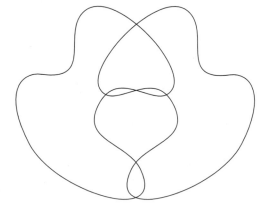

4.

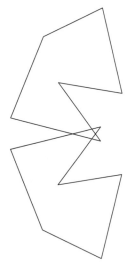

ACE Questions 5–8

5.

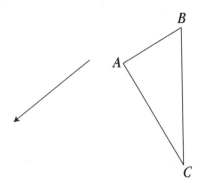

7.

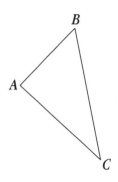

6.

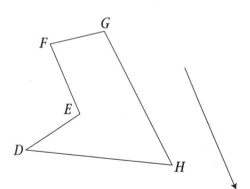

8.

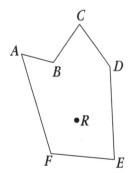

ACE Questions 9–11

9.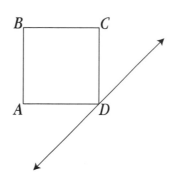

10.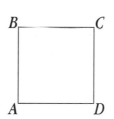

11.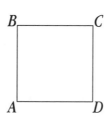

ACE Questions 12–15

12.

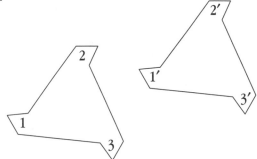

13.

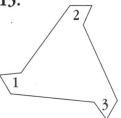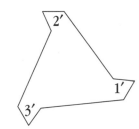

14.

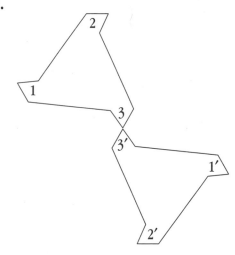

15.

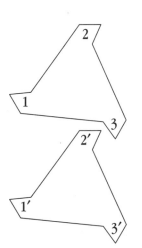

Problem 3.1

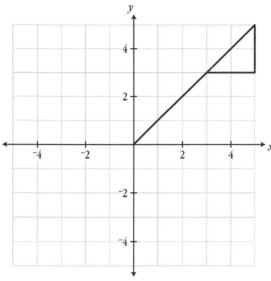

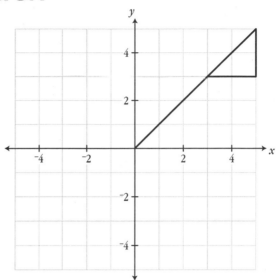

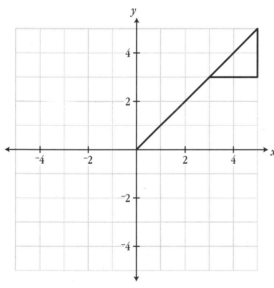

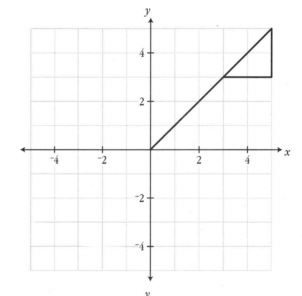

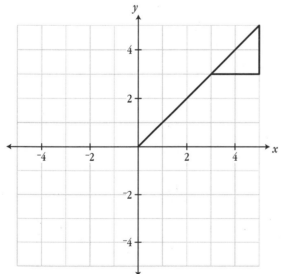

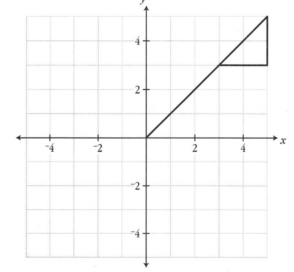

Kaleidoscopes, Hubcaps, and Mirrors

Problem 3.1 Follow-Up

Kaleidoscopes, Hubcaps, and Mirrors

Problem 3.3

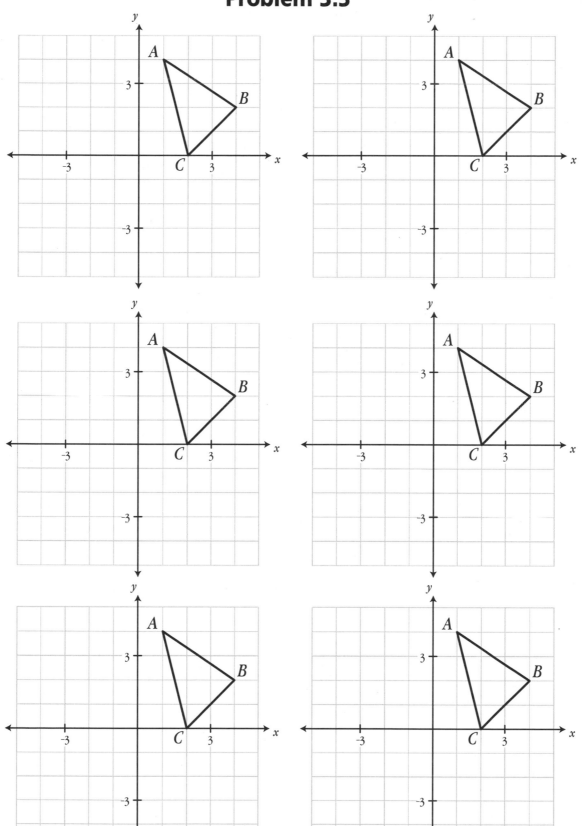

Problem 3.4

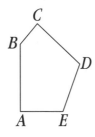
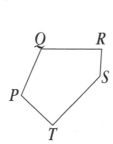
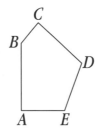
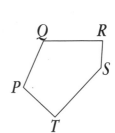

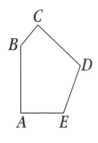
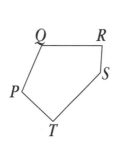
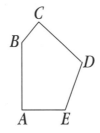
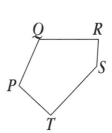

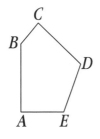
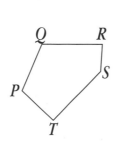
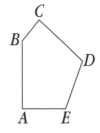
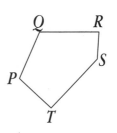

ACE Questions 1–3 and 6–15

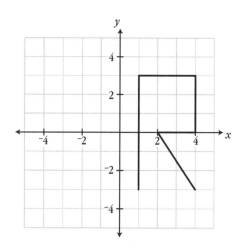

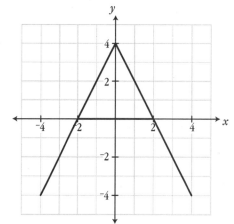

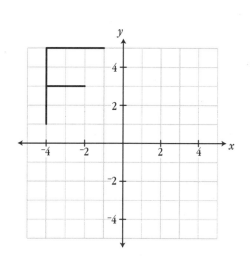

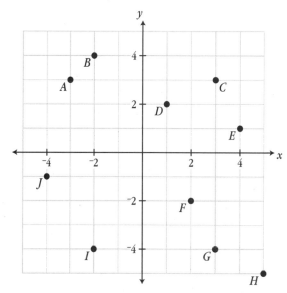

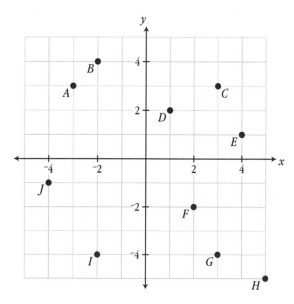

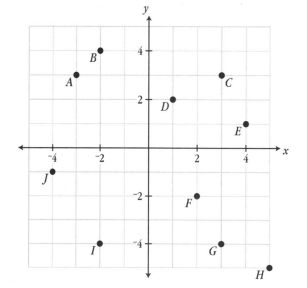

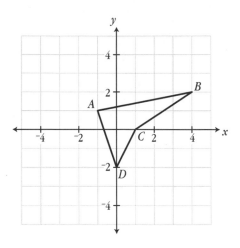

ACE Questions 16 and 24–26

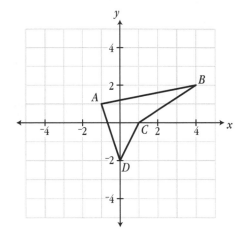

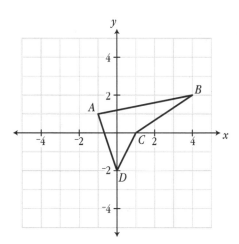

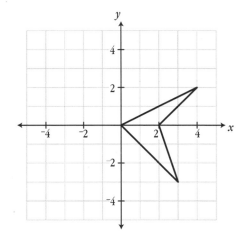

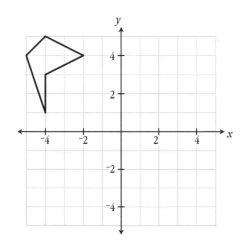

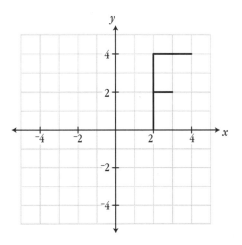

ACE Question 27

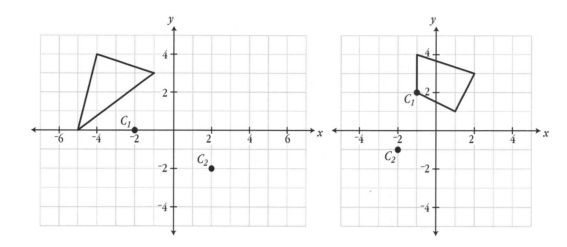

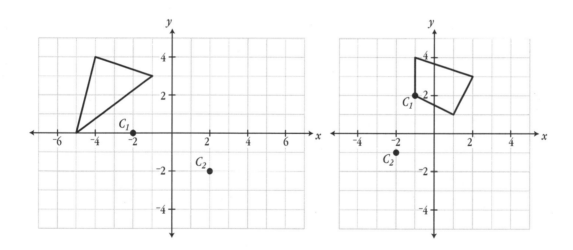

Symmetry Transformations on an Equilateral Triangle

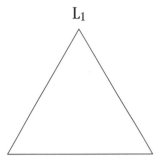

L_1

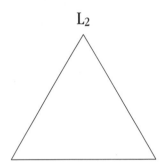

L_2

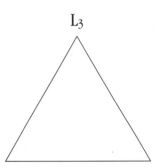

L_3

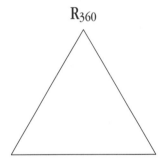

R_{360}

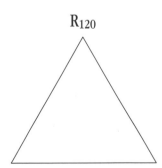

R_{120}

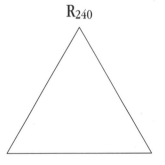

R_{240}

Problem 4.1

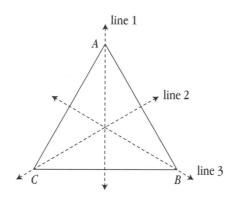

✳	R$_{360}$	R$_{120}$	R$_{240}$	L$_1$	L$_2$	L$_3$
R$_{360}$						
R$_{120}$						
R$_{240}$						
L$_1$					R$_{240}$	
L$_2$						
L$_3$		L$_2$				

×	1	2	3	4	5	6
1						
2						
3						
4						
5						
6						

Symmetry Transformations on a Square

L_1

L_2

L_3

L_4

R_{360}

R_{90}

R_{180}

R_{270}

Problem 4.2

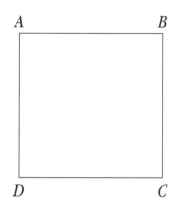

✳	R_{360}							
R_{360}								

×	1	2	3	4	5	6		
1								
2								
3								
4								
5								
6								
7								
8								

ACE Question 8

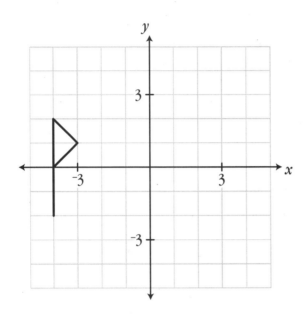

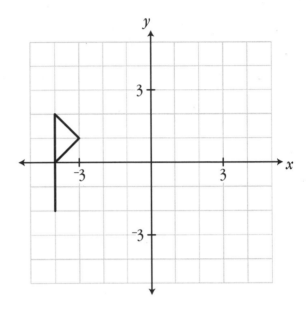

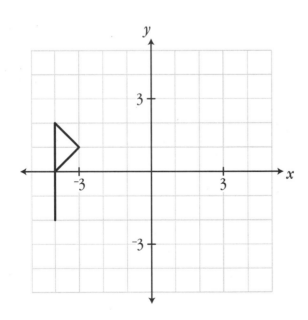

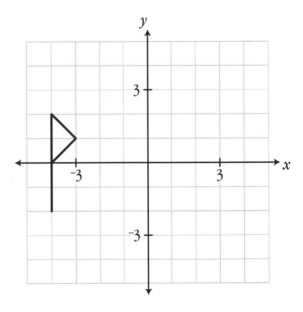

The designs below are reproduced on Labsheet 1.1. Use mirrors, tracing paper, or other tools to help you find all the lines of symmetry in each design.

A.

B.

C.

D.

The pinwheel design below has *rotational symmetry.* It can be rotated 45°, 90°, 135°, 180°, 225°, 270°, 315°, or 360° about its centerpoint to a position in which it looks the same as the original design.

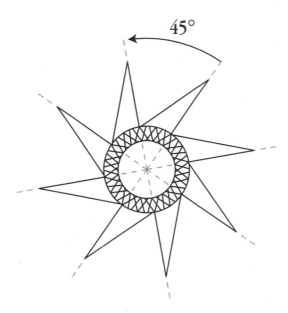

45°

The *angle of rotation* is the smallest angle through which a design can be rotated to coincide with the original design. The angle of rotation for this pinwheel is 45°. Notice that the other rotation angles are multiples of 45°.

Rotational symmetry can be found in many objects that rotate about a centerpoint. For example, the automobile hubcaps shown below have rotational symmetry.

A. Determine the angle of rotation for each hubcap. Explain how you found the angle.

B. Some of the hubcaps also have reflectional symmetry. Sketch all the lines of symmetry for each hubcap.

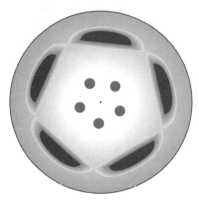

hubcap 1

hubcap 2

hubcap 3

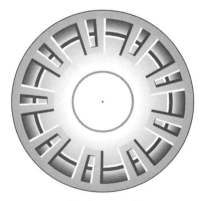

hubcap 4

Do parts A and B for each kaleidoscope design.

A. Look for reflectional symmetry in each design. Sketch all the lines of symmetry you find.

B. Look for rotational symmetry in each design. Determine the angle of rotation for each design.

design 1

design 2

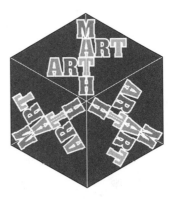

design 3

design 4

design 5

design 6

Each tessellation below has translational symmetry. Do parts A and B for each tessellation.

A. Outline a basic design element that could be used to create the tessellation using only translations.

B. Write directions or draw an arrow showing how the basic design element can be copied and slid to produce another part of the pattern.

9.

10.

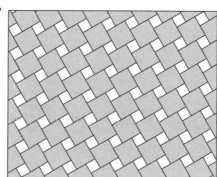

11.

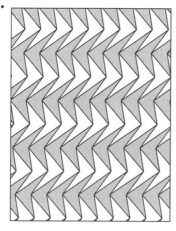

12.

A. Polygon *A′B′C′D′E′* is the image of polygon *ABCDE* under a line reflection.

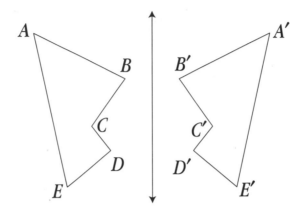

1. Draw a line segment from each vertex of polygon *ABCDE* to its image on polygon *A′B′C′D′E′*.

2. Measure the angles formed by each segment you drew and the line of reflection.

3. For each vertex of polygon *ABCDE*, measure the distance from the vertex to the line of reflection and the distance from the line of reflection to the image of the vertex.

4. Describe the patterns in your measurements from parts 2 and 3.

B. Use what you discovered in part A to draw the image of polygon *JKLMN* under a reflection. Use only a pencil, a ruler, and an angle ruler or protractor. Describe how you drew the image.

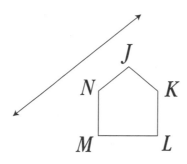

C. Use only a pencil, a ruler, and an angle ruler or protractor to find the line of symmetry for the design below. Describe how you found the line of symmetry.

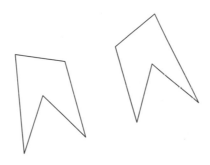

D. Complete this definition of a line reflection: A *line reflection* matches each point *X* on a figure to an image point *X′* so that . . .

A. Each diagram below shows polygon *ABCDE* and its image under a translation. These figures are reproduced on Labsheet 2.2. Do parts 1 and 2 for each diagram.

diagram 1

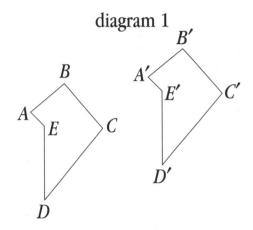

diagram 2

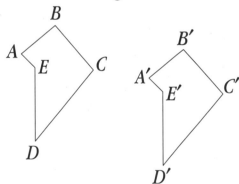

1. Draw a line segment from each vertex of polygon *ABCDE* to its image.

2. Describe the relationship among the line segments you drew.

B. The translations in part A slide polygon *ABCDE* onto its image, polygon *A′B′C′D′E′*. Do parts 1–3 for each diagram in part A.

 1. By performing the same translation that was used to slide polygon *ABCDE* to polygon *A′B′C′D′E′*, slide polygon *A′B′C′D′E′* to create a new image. Label the image *A″B″C″D″E″*.

 2. Polygon *A″B″C″D″E″* is the image of polygon *ABCDE* after two identical translations. How is polygon *A″B″C″D″E″* related to polygon *ABCDE*?

 3. Does your final drawing of the three figures have translational symmetry? Explain.

C. Complete this definition of a translation: A *translation* matches any two points *X* and *Y* on a figure to image points *X′* and *Y′* so that . . .

A. Polygon *A'B'C'D'* is the image of polygon *ABCD* under a rotation of 60° about point *P*. We call point *P* the *center of rotation*.

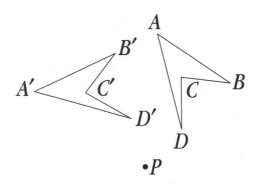

1. What relationship would you expect to find between each vertex, its image, and point *P*?

2. For each vertex of polygon *ABCD*, find the measure of the angle formed by the vertex, point *P*, and the image of the vertex. For example, find the measure of angle *APA'*.

3. For each vertex of polygon *ABCD*, find the distance from the vertex to point *P* and the distance from the image of the vertex to point *P*. For example, find *AP* and *A'P*.

4. What patterns do you see in your measurements? Do these patterns confirm the conjecture you made in part 1?

B. Do parts 1 and 2 for each figure below.

    1. Perform thc indicated rotation, and label the image vertices.

    2. Describe the path each vertex follows under the rotation.

      Rotate 90° about point *P*.   Rotate 45° about point *Q*.

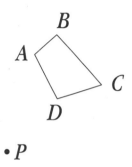 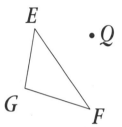

C. For the figures in part B, use the specified rotation to rotate the *image* of the original polygon. The result is the image of the original polygon after two identical rotations. How does the location of the final image compare with the location of the original polygon? Be very specific.

D. Complete this definition of a rotation: A rotation of *d* degrees about a point *P* matches any point *X* on a figure to an image point *X′* so that . . .

A. 1. The figure below is reproduced on Labsheet 2.4A. Reflect triangle *ABC* over line 1. Then reflect the image over line 2. Label the final image *A″B″C″*.

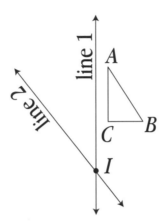

2. For each vertex of triangle *ABC*, measure the angle formed by the vertex, point *I*, and the image of the vertex. For example, measure angle *AIA″*. What do you observe?

3. For each vertex of triangle *ABC*, compare the distance from the vertex to point *I* with the distance from the image of the vertex to point *I*. What do you observe?

4. Could you move triangle *ABC* to triangle *A″B″C″* with a single transformation? If so, describe the transformation.

5. Make a conjecture about the result of reflecting a figure over two intersecting lines. Test your conjecture with an example.

B. 1. What will happen if you reflect a figure over a line and then reflect the image over a second line that is *parallel* to the first line? Would the combination of the two reflections be equivalent to a single transformation?

2. Test your conjecture from part 1 on several examples, including the one shown here. Do the results support your conjecture? If so, explain why. If not, revise your conjecture to better explain your results.

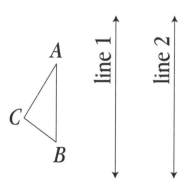

The drawing screen in many computer geometry programs is considered to be a coordinate grid. You can create designs by specifying the endpoints of line segments.

The flag below consists of three segments. The commands for creating the flag tell the computer to draw segments between the specified endpoints.

```
Draw:
Line [(0,-2), (0,3)]
Line [(0,3), (1,2)]
Line [(1,2), (0,1)]
```

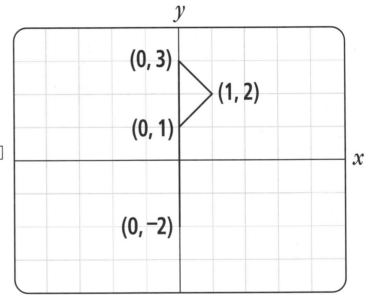

Is there a different set of commands that would create the same flag?

What commands would create a square centered at the origin?

What commands would create a nonsquare rectangle?

A. Copy and complete the commands to create a set of instructions for drawing the flag below.

```
Draw:
Line [( , ), ( , )]
Line [( , ), ( , )]
Line [( , ), ( , )]
```

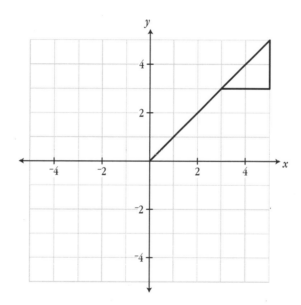

B. Write a set of commands that would draw the image of this flag under a reflection over the *y*-axis.

```
Draw:
Line [( , ), ( , )]
Line [( , ), ( , )]
Line [( , ), ( , )]
```

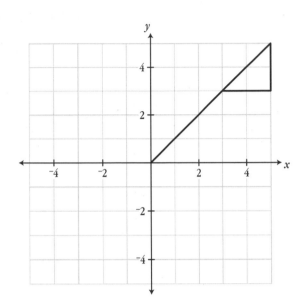

C. Write a set of commands that would draw the image of this flag under a reflection over the *x*-axis.

```
Draw:
Line [(   ,   ), (   ,   )]
Line [(   ,   ), (   ,   )]
Line [(   ,   ), (   ,   )]
```

D. Write a set of commands that would draw the image of this flag under a reflection over the line $y = x$.

```
Draw:
Line [(   ,   ), (   ,   )]
Line [(   ,   ), (   ,   )]
Line [(   ,   ), (   ,   )]
```

A. 1. In diagram 1, the left-most flag can be drawn with these commands:

```
Draw:
Line [(-5,-4), (-5,2)]
Line [(-5,2), (-4,1)]
Line [(-4,1), (-5,0)]
```

This draws the vertical segment, then the upper slanted segment, and finally the lower slanted segment.

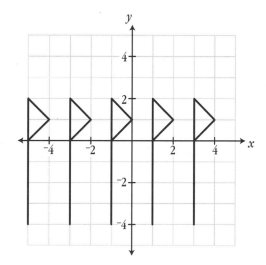

Write sets of commands for drawing the other four flags in diagram 1. Each set of commands should draw the segments in the same order as the commands for the original flag.

Flag 2
```
Draw:
Line [( , ), ( , )]
Line [( , ), ( , )]
Line [( , ), ( , )]
```

Flag 3
```
Draw:
Line [( , ), ( , )]
Line [( , ), ( , )]
Line [( , ), ( , )]
```

Flag 4
```
Draw:
Line [( , ), ( , )]
Line [( , ), ( , )]
Line [( , ), ( , )]
```

Flag 5
```
Draw:
Line [( , ), ( , )]
Line [( , ), ( , )]
Line [( , ), ( , )]
```

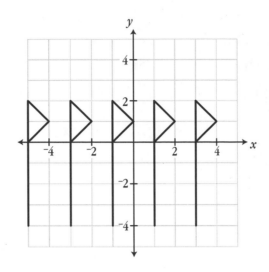

A. 2. Compare the commands for the five flags. Describe a pattern that relates the coordinates of each flag to the coordinates of the flag to its *right*.

```
Draw:
Line [(   ,   ), (   ,   )]
Line [(   ,   ), (   ,   )]
Line [(   ,   ), (   ,   )]
```

3. Describe a pattern that relates the coordinates of each flag to the coordinates of the flag to its *left*.

```
Draw:
Line [(   ,   ), (   ,   )]
Line [(   ,   ), (   ,   )]
Line [(   ,   ), (   ,   )]
```

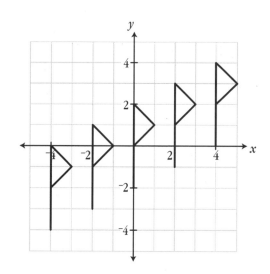

B. 1. Write a set of commands for drawing the left-most flag in diagram 2. Then write comparable instructions for drawing the other four flags.

Flag 1
Draw:
Line [(,), (,)]
Line [(,), (,)]
Line [(,), (,)]

Flag 2
Draw:
Line [(,), (,)]
Line [(,), (,)]
Line [(,), (,)]

Flag 3
Draw:
Line [(,), (,)]
Line [(,), (,)]
Line [(,), (,)]

Flag 4
Draw:
Line [(,), (,)]
Line [(,), (,)]
Line [(,), (,)]

Flag 5
Draw:
Line [(,), (,)]
Line [(,), (,)]
Line [(,), (,)]

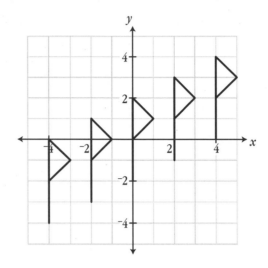

B. 2. Compare the commands for the five flags. Describe a pattern that relates the coordinates of each flag to the coordinates of the flag to its *right*.

3. Describe a pattern that relates the coordinates of each flag to the coordinates of the flag to its *left*.

A. Copy and complete the commands to create a set of instructions for drawing triangle *ABC*.

```
Draw:
Line [( , ), ( , )]
Line [( , ), ( , )]
Line [( , ), ( , )]
```

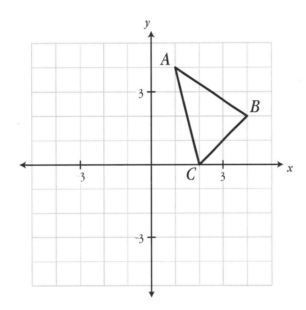

B. Write a set of commands that would draw the image of triangle *ABC* under a 90° rotation about the origin.

```
Draw:
Line [( , ), ( , )]
Line [( , ), ( , )]
Line [( , ), ( , )]
```

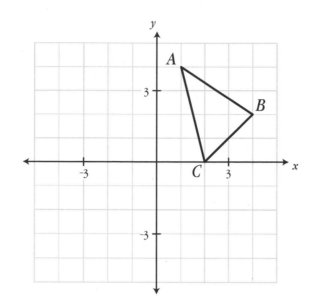

C. Write a set of commands that would draw the image of triangle *ABC* under a 180° rotation about the origin.

```
Draw:
Line [(  ,  ), (  ,  )]
Line [(  ,  ), (  ,  )]
Line [(  ,  ), (  ,  )]
```

D. Write a set of commands that would draw the image of triangle *ABC* under a 270° rotation about the origin.

```
Draw:
Line [(  ,  ), (  ,  )]
Line [(  ,  ), (  ,  )]
Line [(  ,  ), (  ,  )]
```

E. Write a set of commands that would draw the image of triangle *ABC* under a 360° rotation about the origin.

```
Draw:
Line [(  ,  ), (  ,  )]
Line [(  ,  ), (  ,  )]
Line [(  ,  ), (  ,  )]
```

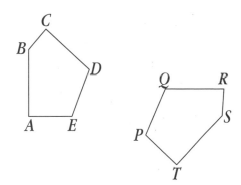

A. If you made a copy of one of the pentagons and fit it exactly on the other, which vertices would match?

B. Which pairs of sides in pentagons *ABCDE* and *PQRST* are the same length?

C. Which pairs of angles in pentagons *ABCDE* and *PQRST* are the same size?

D. What combination of reflections, rotations, and translations would move pentagon *ABCDE* to fit exactly on pentagon *PQRST*? Is there more than one possible combination? Make sketches to show your ideas.

Complete the table to show the results of combining symmetry transformations of an equilateral triangle. Each entry should be the result of performing the transformation in the left column followed by the transformation in the top row.

The two entries already in the table represent the combinations you explored in the introduction:

$$L_1 * L_2 = R_{240} \qquad L_3 * R_{120} = L_2$$

$*$	R_{360}	R_{120}	R_{240}	L_1	L_2	L_3
R_{360}						
R_{120}						
R_{240}						
L_1					R_{240}	
L_2						
L_3		L_2				

Note that transformation R_{360}, a 360° rotation, carries every point back to where it started. As you combine transformations, you will discover that many combinations are equivalent to R_{360}.

A. On Labsheet 4.2B, draw all the lines of symmetry and describe all the rotations that produce images that exactly match the original square. Label the lines of symmetry in clockwise order as line 1, line 2, and so on.

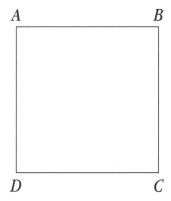

B. Cut out a copy of the square, and copy each vertex label onto the back side of the square. Use this copy to explore combinations of symmetry transformations. Complete the operation table to show the results of combining pairs of transformations. When you enter the column heads, first list the rotations and then the reflections. Use the same order for the row heads.

$*$	R_{360}							
R_{360}								

The operations of addition and multiplication satisfy important properties that are useful for reasoning about expressions and equations.

- The order in which numbers are added or multiplied does not affect the result. This is called the commutative property. We say that addition and multiplication are *commutative operations.* In symbols, if a and b are real numbers, then

$$a + b = b + a \quad \text{and} \quad a \times b = b \times a.$$

- Adding 0 to a number has no effect. Multiplying a number by 1 has no effect. In symbols, if a is a real number, then

$$0 + a = a + 0 = a \quad \text{and} \quad 1 \times a = a \times 1 = a.$$

We call 0 and 1 *identity elements;* 0 is the additive identity, and 1 is the multiplicative identity.

- For any number a, the sum of a and $-a$ is 0, the additive identity. For any nonzero number a, the product of a and $\frac{1}{a}$ is 1, the multiplicative identity.

$$a + {-a} = {-a} + a = 0 \quad \text{and} \quad a \times \frac{1}{a} = \frac{1}{a} \times a = 1$$

We call $-a$ the *additive inverse* of a and $\frac{1}{a}$ the *multiplicative inverse* of a.

Refer to your operation table for combining symmetry transformations for an equilateral triangle.

A. Is ✳ a commutative operation? In other words, does the order in which you combine symmetry transformations make a difference? Justify your answer.

B. Like addition and multiplication, the ✳ operation has an identity element. That is, there is a transformation that has no effect when it is applied before or after another transformation. Tell which transformation is the identity element, and explain how you know.

C. Does each symmetry transformation have an inverse? That is, can you combine each transformation with another transformation, in either order, to get the identity element you found in part B? If so, list each transformation and its inverse.

Dear Family,

We can see symmetry all around us, in designs on gift wrap and fabrics and pottery. Symmetry can be over a line, as we see in a kite or a butterfly. Symmetry can be about a point, as in pinwheels and hubcaps. And it can be in a strip pattern such as those often seen on wallpaper borders. The late Dutch artist M. C. Escher (1898–1972) is famous for the use of symmetry in his artwork.

The next unit your child will be studying in mathematics class this year involves the geometry of symmetry and the kinds of motion that can be used to create symmetrical designs. *Kaleidoscopes, Hubcaps, and Mirrors* is an introduction to the topic in mathematics called *transformational geometry.*

We often think of algebra and geometry as two different branches of mathematics. In this unit, students will see some of the many ways in which algebra and geometry complement and reinforce each other.

Here are some strategies for helping your child during this unit:

- Talk with your child about the ideas presented in the text about symmetry. Look with your child for examples of each type of symmetry.

- Talk with your child about careers that use the knowledge of geometry, such as crystallography, a science that deals with the forms and structures of crystals.

- Encourage your child's efforts in completing all homework assignments.

As always, if you have any questions or suggestions about your child's mathematics program, please feel free to call. We are interested in your child and want to be sure this year's mathematics experiences are enjoyable and promote a firm understanding of mathematics.

Sincerely,

Estimada familia,

La simetría está presente en todo nuestro entorno, en los dibujos que aparecen en el papel de envolver regalos, en los tejidos y en la cerámica. La simetría puede presentarse con respecto a una recta, como ocurre con las cometas o las mariposas. También puede darse con respecto a un punto, como es el caso de los molinetes y los tapacubos. E incluso puede presentarse la simetría en los patrones de las bandas decorativas como, por ejemplo, en los que aparecen frecuentemente en los frisos del papel de empapelar. El fallecido artista holandés M. C. Escher (1898–1972) es famoso por utilizar en sus obras la simetría.

La próxima unidad que su hijo o hija estudiará este año en la clase de matemáticas trata sobre la geometría de la simetría y sobre los tipos de movimientos que pueden emplearse para crear diseños simétricos. *Kaleidoscopes, Hubcaps, and Mirrors* (Calidoscopios, tapacubos, y espejos) sirve de introducción al tema matemático llamado *geometría de transformaciones.*

Muchas veces consideramos al álgebra y a la geometría como a dos ramas diferentes de las matemáticas. En esta unidad los alumnos conocerán algunas de las diversas maneras en que el álgebra y la geometría se complementan y se refuerzan mutuamente.

He aquí algunas estrategias que ustedes pueden emplear para ayudar a su hijo o hija en esta unidad:

- Comenten con él o ella las ideas que se presentan en el texto en relación a la simetría. Busquen ejemplos de cada uno de los tipos de simetría.

- Hablen juntos sobre las profesiones en las que se emplean conocimientos geométricos, como puede ser en la cristalografía, la ciencia que estudia las formas y las estructuras de los cristales.

- Animen a su hijo o hija a esforzarse para que complete toda la tarea.

Y como de costumbre, si ustedes tienen alguna duda o recomendación relacionada con el programa de matemáticas de su hijo o hija, no duden en llamarnos. Nos interesa su hijo o hija y queremos asegurarnos de que las experiencias matemáticas que tenga este año sean lo más amenas posibles y ayuden a fomentar en él o ella una sólida comprensión de las matemáticas.

Atentamente,

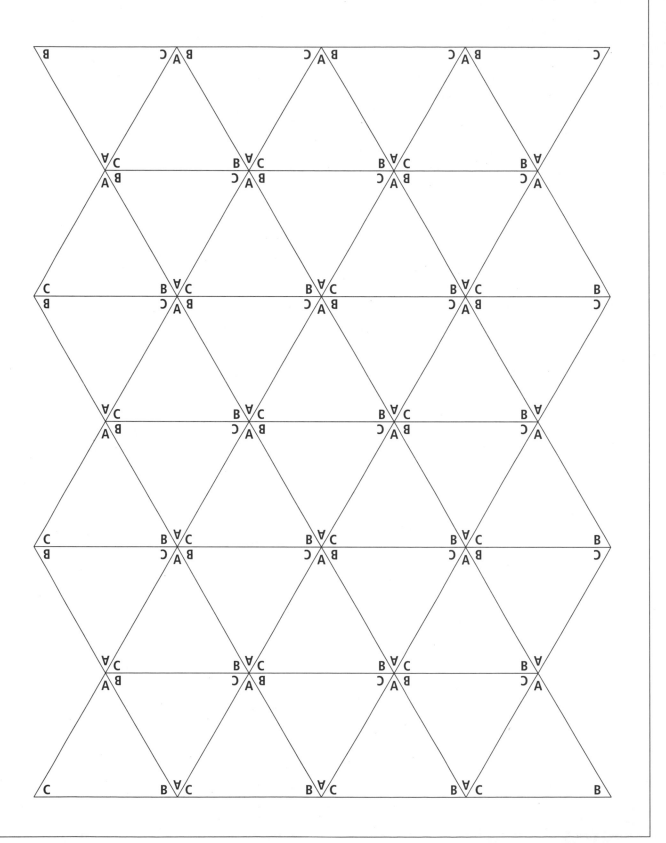

A	B	A	B	A	B	A	B
C	D	C	D	C	D	C	D
A	B	A	B	A	B	A	B
C	D	C	D	C	D	C	D
A	B	A	B	A	B	A	B
C	D	C	D	C	D	C	D
A	B	A	B	A	B	A	B
C	D	C	D	C	D	C	D
A	B	A	B	A	B	A	B
C	D	C	D	C	D	C	D

Hexagon *ABCDEF*

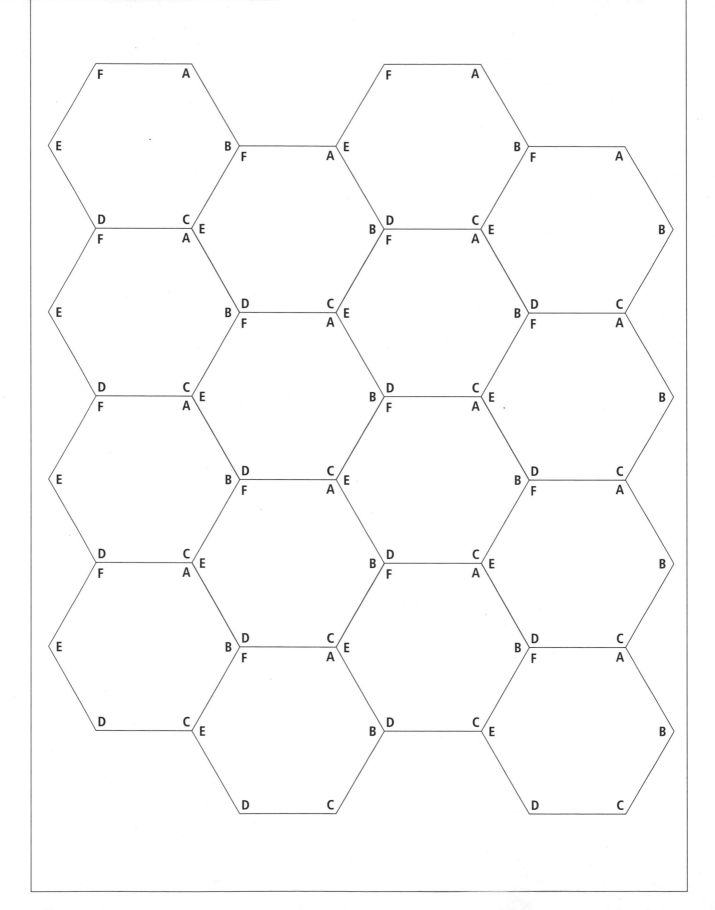

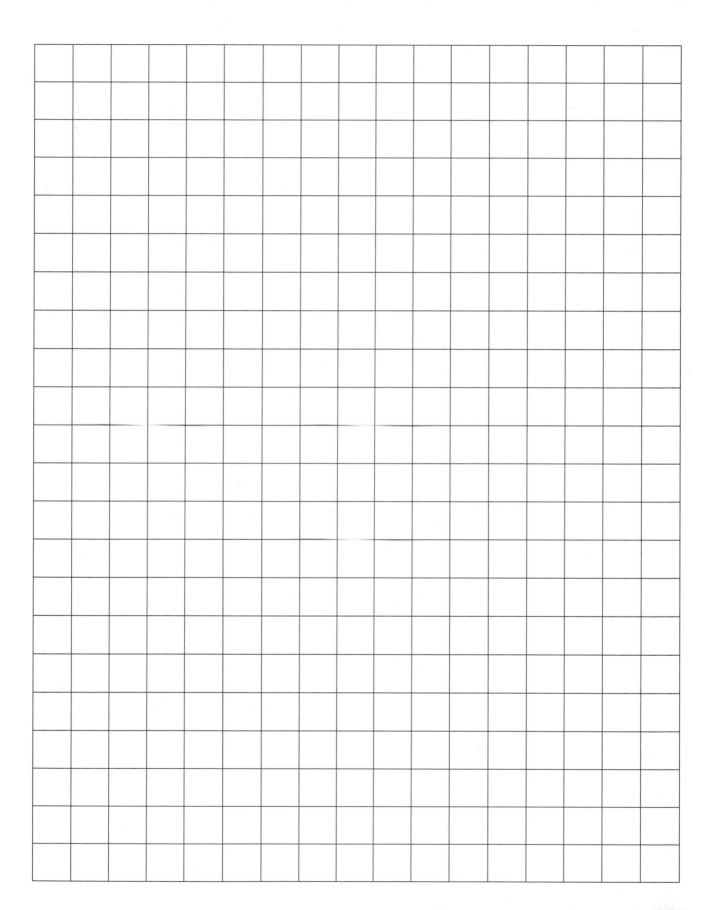

Quarter-Inch Grid Paper

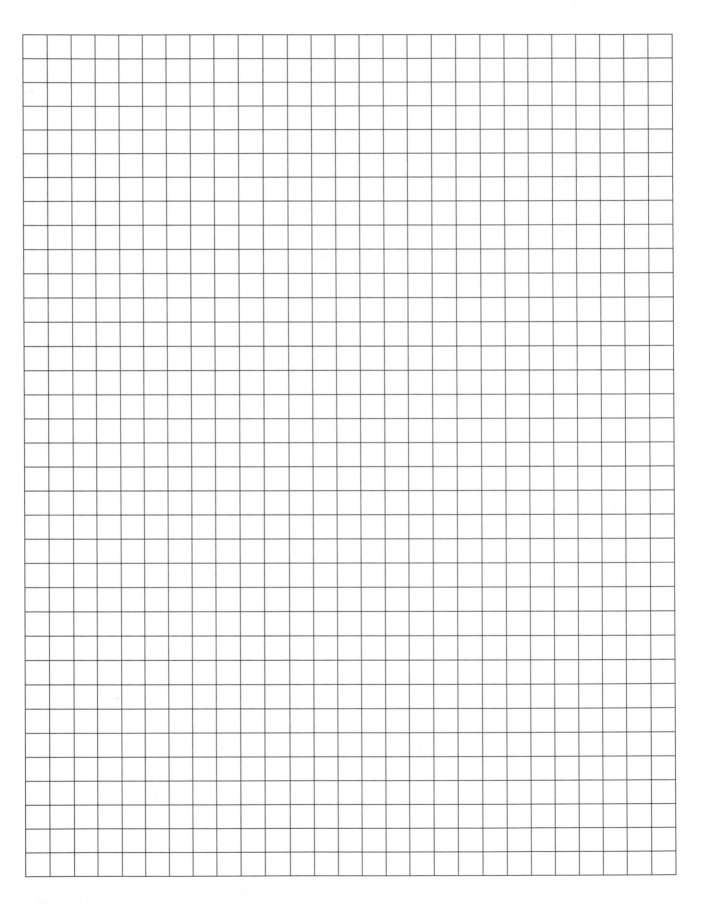

Additional Practice

Investigation 1

Use these problems for additional practice after Investigation 1.

In 1–6, determine all the types of symmetry in the design. Specify lines of symmetry, centers and angles of rotation, and lengths and directions of translations.

1.

2.

3.
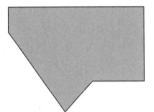

4.
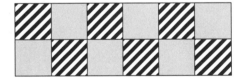

5.
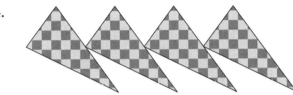

6.
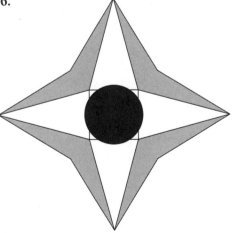

In 7 and 8, a basic design element and one or more lines are given. Use the basic design element to create a design with the given lines as lines of symmetry.

7.

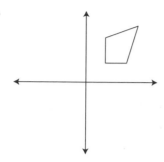

8.

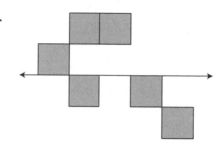

In 9 and 10, indicate the lines of symmetry and the center and angle of rotation for the line design.

9.

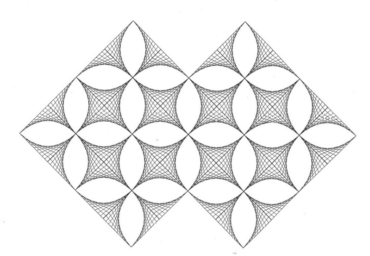

10.

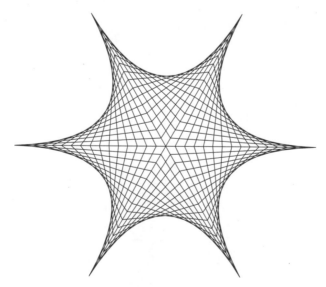

Source: Dale Seymour. *Introduction to Line Designs*. Palo Alto, Calif.: Dale Seymour Publications, 1992.

Investigation 2

Use these problems for additional practice after Investigation 2.

In 1 and 2, describe a reflection or a combination of two reflections that would move shape 1 to exactly match shape 2.

1.

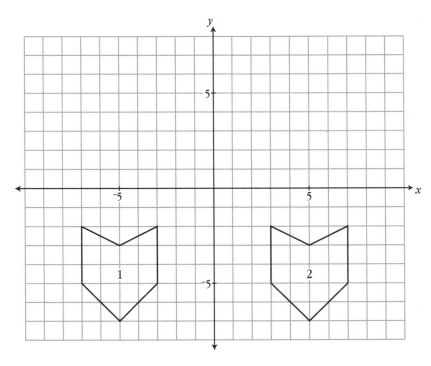

2.

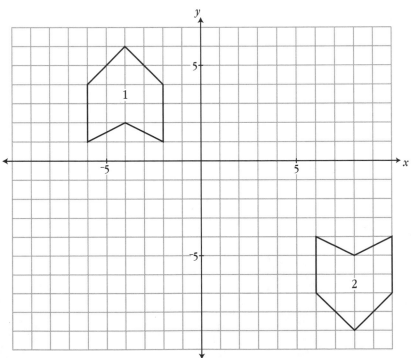

In 3 and 4, draw the image of the polygon under a reflection over the line. Describe what happens to each point on the original polygon under the reflection.

3.

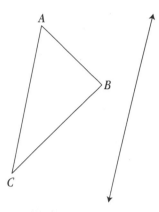

4.

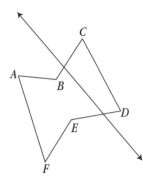

In 5 and 6, a shape and its image under a line reflection are given. Do parts a and b.

5.

6.

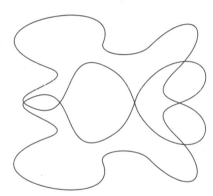

a. Draw the line of symmetry for the figure.

b. Label three points on the figure, and label the corresponding image points.

In 7 and 8, perform the translation indicated by the arrow. Describe what happens to each point of the original figure under the translation.

7.

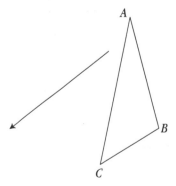

8.

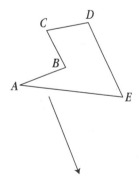

9. Rotate triangle *ABC* 90° about point *R*. Describe what happens to each point of triangle *ABC* under the rotation.

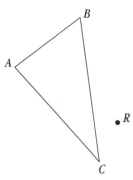

10. Rotate polygon *ABCDEF* 180° about point *F*. Describe what happens to each point of polygon *ABCDEF* under the rotation.

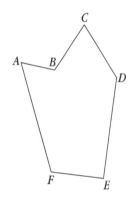

In 11–13, refer to this diagram.

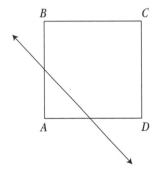

11. Draw the image of square *ABCD* under a reflection over the line.

12. Draw the image of square *ABCD* under a 45° rotation about point *D*.

13. Draw the image of square *ABCD* under the translation that slides point *D* to point *P*.

In 14–17, a polygon and its image under a transformation are given. Decide whether the transformation was a line reflection, a rotation, or a translation. Then indicate the reflection line, the center and angle of rotation, or the direction and distance of the translation.

14.

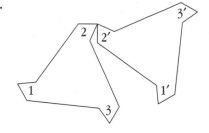

15.

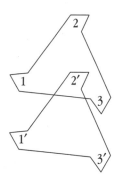

16.

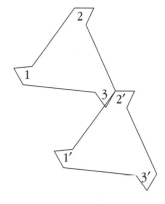

17.

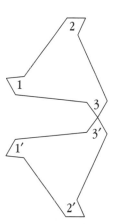

Investigation 3

Use these problems for additional practice after Investigation 3.

In 1 and 2, assume that the pattern in the graph continues in both directions. Identify a basic design element that could be copied and transformed to create the entire pattern, and describe how the pattern could be created from that design element.

1.

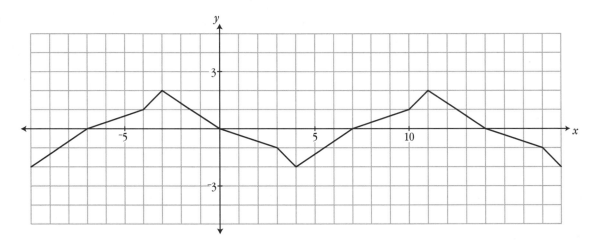

2.

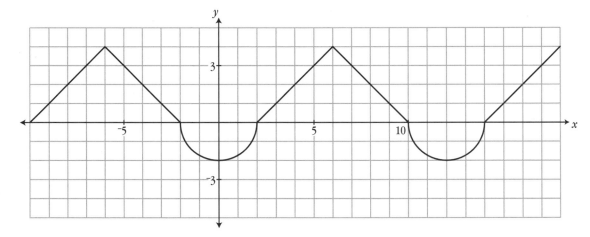

3. Plot the points (2, 4), (3, 5), (5, 5), (4, 4), (5, 3), and (3, 3) on a coordinate grid. Form a polygon by connecting the points in order and then connecting the last point to the first point. Reflect the polygon over the *y*-axis. Then translate the image 6 units to the right. Finally, rotate the second image 90° about the origin. What are the coordinates of the vertices of the final image?

4. Suppose the shape below is translated according to the rolls of a six-sided number cube.
   - If a 1, 2, or 3 is rolled, the shape is translated 3 units to the right.
   - If a 4 is rolled, the shape is translated 3 units up.
   - If a 5 is rolled, the shape is translated 3 units down.
   - If a 6 is rolled, the shape is translated 3 units to the left.

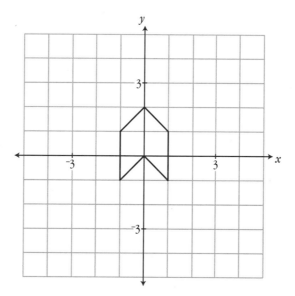

   a. Draw the shape in its location after the following sequence of rolls: 3, 5, 6. What are the new coordinates of a general point (x, y) on the shape after this sequence of rolls?

   b. Draw the shape in its location after the following sequence of rolls: 1, 6, 4, 2. What are the new coordinates of a general point (x, y) on the shape after this sequence of rolls?

   c. What sequence of rolls will produce a final image whose coordinates are all negative?

5. Describe two different sets of transformations that would move square *PQRS* onto square *WXYZ*.

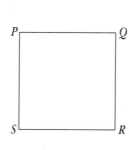 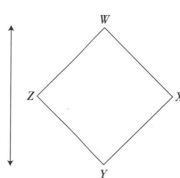

In 6–11, refer to the grid below.

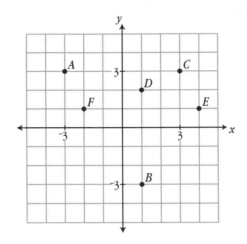

6. What are the coordinates of the image of point *A* under a translation that moves point (1, 2) onto point (⁻2, 0)?

7. What are the coordinates of the image of point *B* under a translation that moves point (1, 2) onto point (4, ⁻4)?

8. What are the coordinates of the image of point *C* under a translation that moves point (1, 2) onto point (⁻3, ⁻2)?

9. What are the coordinates of the image of point *D* under a reflection over the *x*-axis?

10. What are the coordinates of the image of point *E* under a reflection over the *y*-axis?

11. What are the coordinates of the image of point *F* under a reflection over the line *y* = *x*?

12. Identify two congruent shapes in the figure below, and explain how you could use symmetry transformations to move one shape onto the other.

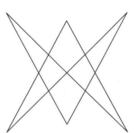

In 13 and 14, refer to the grid at right.

13. Describe how you could move shape 1 to exactly match shape 1′ by using at least one translation and at least one reflection.

14. Describe how you could move shape 2 to exactly match shape 2′ by using at least one translation and at least one reflection.

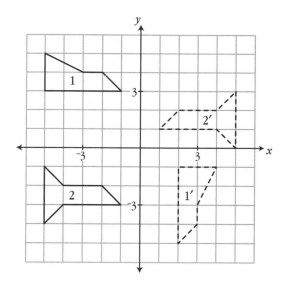

In 15–17, refer to the grid below.

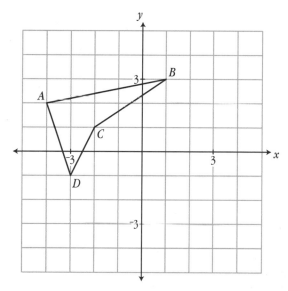

15. **a.** Draw the final image created by rotating polygon *ABCD* 90° about the origin and then reflecting the image over the *x*-axis.

b. Draw the final image created by reflecting polygon *ABCD* over the *x*-axis and then rotating the image 90° about the origin.

c. Are the final images in parts a and b the same? Why or why not?

16. What single transformation is equivalent to a rotation of 90° about the origin followed by a rotation of 270° about the origin?

17. What single transformation is equivalent to a reflection over the *y*-axis, followed by a reflection over the *x*-axis, followed by a reflection over the *y*-axis?

Investigation 4

Use these problems for additional practice after Investigation 4.

1. Refer to the grid below.

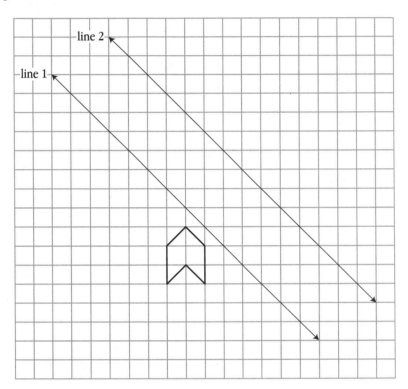

a. Reflect the shape over line 1, and then reflect the image over line 2. Describe a single transformation that would give the same final image.

b. Reflect the shape over line 2, and then reflect the image over line 1. Describe a single transformation that would give the same final image.

c. Is reflecting figures over parallel lines a commutative operation? Explain your answer.

2. Describe all the symmetry transformations for rhombus *ABCD*. Then make a table showing the results of all possible combinations of two transformations.

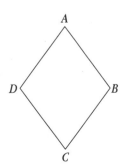

3. Shakaya has developed a system for combining pairs of letters in the set {a, b, c, d}. She uses a ♥ symbol to represent her combining operation. She made the table below to show the result of each combination of letters.

♥	a	b	c	d
a	b	c	d	a
b	c	d	a	b
c	d	a	b	c
d	a	b	c	d

a. Is Shakaya's operation commutative? Explain.

b. What is the identity element?

c. For each letter, give the inverse and explain how you know it is the inverse.

d. Evaluate each expression.

  i. (b ♥ c) ♥ a

  ii. (a ♥ b) ♥ (b ♥ c)

Answer Keys

Investigation 1

1. The design has reflectional symmetry over the lines shown and rotational symmetry with a 180° angle of rotation about point *P.*

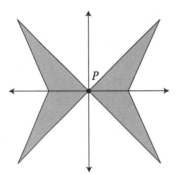

2. The design has reflectional symmetry over the line shown.

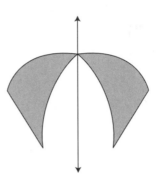

3. This design has no symmetries.

4. This design has translational symmetry with length and direction indicated by the arrow.

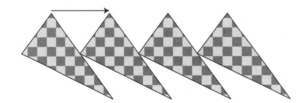

5. This design has rotational symmetry with a 180° angle of rotation about point *P* and translational symmetry with length and direction indicated by the arrow.

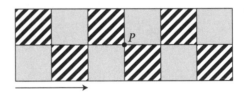

6. This design has rotational symmetry with a 90° angle of rotation about point *P* and reflectional symmetry over the lines shown.

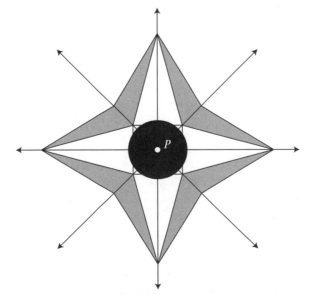

7.

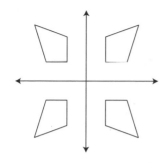

8.

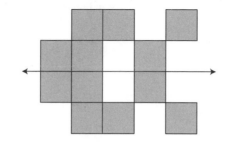

9. The design has a vertical line of symmetry through its center and a horizontal line of symmetry through its center. It has rotational symmetry with a 180° angle of rotation about its centerpoint.

10. The design has six lines of symmetry. Each line through opposite star points is a line of symmetry, and each line through opposite "turning points" of the curved sections is a line of symmetry. The design has rotational symmetry with a 60° angle of rotation about the centerpoint.

Investigation 2

1. a reflection over the *y*-axis

2. Possible answer: a reflection over the line $x = 2$, followed by a reflection over the line $y = -1.5$

In 3 and 4, each point is matched to an image point on the other side of the line of reflection. The image point lies on the line passing through the original point, perpendicular to the line of reflection. The distance from the image point to the line of reflection is equal to the distance from the original point to the line of reflection.

3.

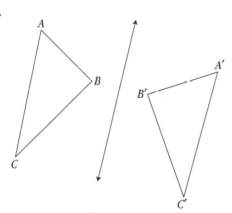

4.

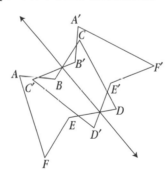

5. Possible answer:

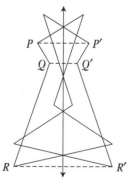

6. Possible answer:

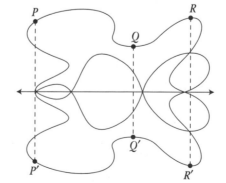

In 7 and 8, each point on the original figure is matched to an image point whose distance and direction from the original point are determined by the arrow.

7.

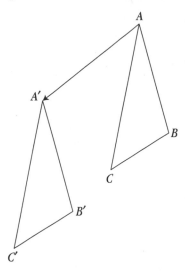

8.

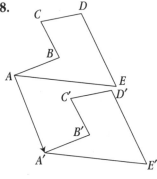

9. Each point *X* on triangle *ABC* is matched to an image point *X′* so that the $RX = RX′$ and the measure of angle *XRX′* is 90°.

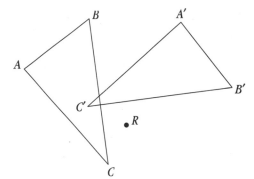

10. Each point *X* on polygon *ABCDEF* is matched to an image point *X′* so that the $FX = FX′$ and the measure of angle *XFX′* is 180°.

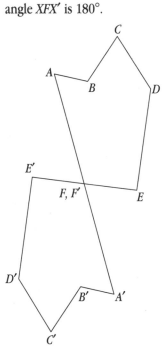

11.

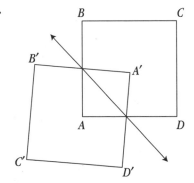

12.

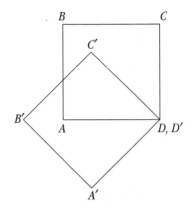

13.

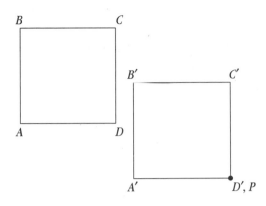

14. a rotation of 90° about point *P*

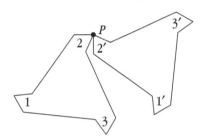

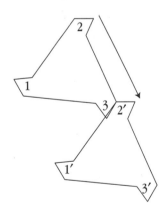

15. a translation with the length and direction indicated by the arrow

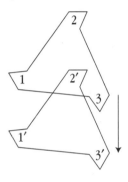

16. a translation with the length and direction indicated by the arrow

17. a reflection over the line shown

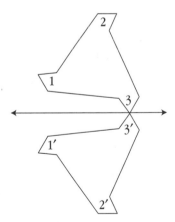

Investigation 3

1. One possible basic design element is shown below. You can generate the entire pattern by repeatedly translating this element 7 units to the right or left and then reflecting alternate elements over the *x*-axis.

2. One possible basic design element is shown below. If you start with the copy of this element immediately to the left of the *y*-axis, you can generate the right half of the pattern by reflecting this element over the line *x* = 0, then *x* = 6, then *x* = 12, then *x* = 18, and so on. You can generate the left half of the pattern by reflecting this element over the line *x* = −6, then *x* = −12, then *x* = −18, and so on.

3. The coordinates of the final image are (−4, 4), (−5, 3), (−5, 1), (−4, 2), (−3, 1), and (−3, 3).

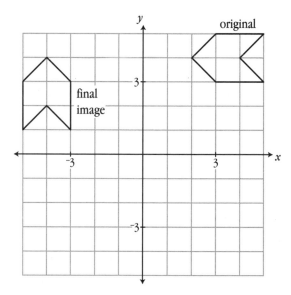

4. **a.** The result is labeled *shape a* in the drawing below. The coordinates of a general point (x, y) after the sequence of rolls is $(x, y - 3)$.

   b. The result is labeled *shape b* in the drawing below. The coordinates of a general point (x, y) after the sequence of rolls is $(x + 3, y + 3)$.

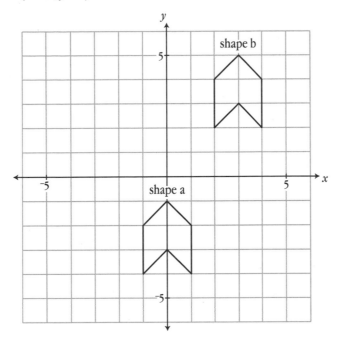

   c. Possible answer: 5, 6

5. Possible answer: Reflect square *PQRS* over the line and then rotate it 45° about its centerpoint. Rotate square *PQRS* 135° and then translate it until vertex *P* matches vertex *Y*.

6. $(^-6, 1)$ 7. $(4, ^-9)$ 8. $(^-1, ^-1)$

9. $(1, ^-2)$ 10. $(^-4, 1)$ 11. $(1, ^-2)$

12. Possible answer: One of the congruent shapes is drawn with a dashed line and the other with a solid line. You can move one shape onto the other by reflecting it over the line shown.

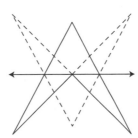

13. Possible answer: Translate shape 1 down 1 unit and to the left 4 units. Then reflect it over the line $x = ^-5$. Finally, reflect it over the line $y = x$.

14. Possible answer: Reflect shape 2 over the *y*-axis and then translate it up 4 units.

15. a. The final image is labeled *image a*.

b. The final image is labeled *image b*.

c. The images are not the same. Rotating a figure 90° about the origin and then reflecting it over the *x*-axis takes point (x, y) to $(^-y, x)$ and then to $(^-y, ^-x)$. Reflecting a figure over the *x*-axis and then rotating it 90° about the origin takes point (x, y) to $(x, ^-y)$ and then to (y, x).

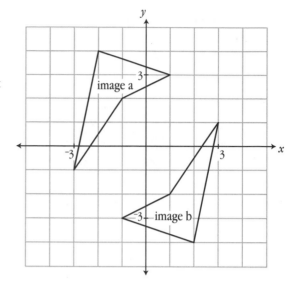

16. a 360° rotation about the origin

17. a reflection over the *x*-axis

Investigation 4

1. a. The image is labeled *image a*. You would get the same image by translating the original figure according to the rule $(x, y) \rightarrow (x + 5, y + 5)$.

b. The image is labeled *image b*. You would get the same image if you translated the original figure according to the rule $(x, y) \rightarrow (x - 5, y - 5)$.

c. No. A reflection over line 1 followed by a reflection over line 2 did not give the same image as a reflection over line 2 followed by a reflection over line 1.

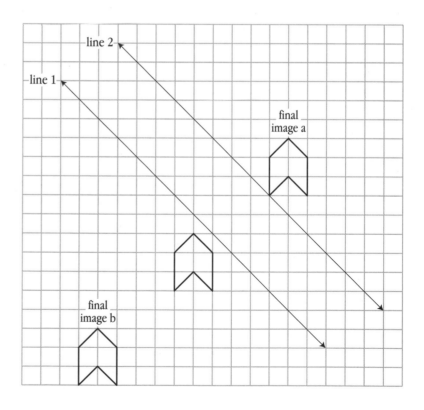

2. The symmetry transformations are L_1, L_2, R_{180}, R_{360}, where L_1 is a reflection over line 1 and L_2 is a reflection over line 2.

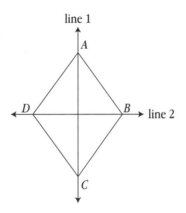

line 1

A

D B line 2

C

Below is the table of all possible combinations.

$*$	R_{360}	R_{180}	L_2	L_1
R_{360}	R_{360}	R_{180}	L_2	L_1
R_{180}	R_{180}	R_{360}	L_1	L_2
L_2	L_2	L_1	R_{360}	R_{180}
L_1	L_1	L_2	R_{180}	R_{360}

3. **a.** Yes. The order in which elements are combined does not matter. In the table, this is illustrated by the symmetry of the elements about the diagonal from the upper left to the lower right.

   b. d

   c. The letters a and c are inverses since a ♥ c = c ♥ a = d. The letter b is its own inverse since b ♥ b = d. The letter d is its own inverse since d ♥ d = d.

   d. **i.** (b ♥ c) ♥ a = a ♥ a = b

   ii. (a ♥ b) ♥ (b ♥ c) = c ♥ a = d

congruent figures Two figures are congruent if one is an image of the other under a translation, a reflection, a rotation, or some combination of these transformations. Put more simply, two figures are congruent if you can slide, flip, or turn one figure so that it fits exactly on the other. The polygons below are congruent.

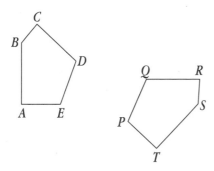

kaleidoscope A tube containing colored beads or pieces of glass and carefully placed mirrors. When a kaleidoscope is held to the eye and rotated, the viewer sees colorful, symmetric patterns.

line reflection A transformation that matches each point on a figure with its mirror image over a line. Polygon *A'B'C'D'E'* below is the image of polygon *ABCDE* under a reflection over the line. If you drew a line segment from a point to its image, the segment would be perpendicular to and bisected by the line of reflection.

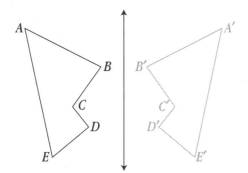

reflectional symmetry A figure or design has reflectional symmetry if you can draw a line that divides the figure into halves that are mirror images. The line that divides the figure into halves is called the line of symmetry. The figure below has reflectional symmetry about a vertical line through its center.

rotation A transformation that turns a figure counterclockwise about a point. Polygon *A'B'C'D'* below is the image of polygon *ABCD* under a 60° rotation about point *P*. If you drew a segment from a point on polygon *ABCD* to point *P* and another segment from the point's image to point *P*, the segments would be the same length and they would form a 60° angle.

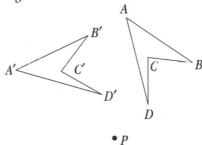

rotational symmetry A figure or design has rotational symmetry if it can be rotated less than a full turn about a point to a position in which it looks the same as the original. The hubcap design below has rotational symmetry with its center as the center of rotation and a 120° angle of rotation. This means that it can be rotated 120°, or any multiple of 120°, about its centerpoint to create an image that matches exactly with the original.

symmetry   An object or design has symmetry if part of it is repeated to create a balanced pattern. In this unit, students learn about three types of symmetry. The butterfly below has reflectional symmetry, the fan blades have rotational symmetry, and the wallpaper design has translational symmetry.

tessellation   A design made from copies of a basic design element that cover a surface without gaps or overlaps. Tessellations have translational symmetry. The design below is a tessellation.

transformation   A geometric operation that matches each point on a figure with an image point. A symmetry transformation produces an image that is identical in size and shape to the original figure. Reflections, rotations, and translations are types of symmetry transformations.

translation   A transformation that slides each point on a figure to an image point a given distance and direction from the original point. Polygon *A′B′C′D′E′* below is the image of polygon *ABCDE* under a translation. If you drew line segments from two points to their respective image points, the segments would be parallel and they would have the same length.

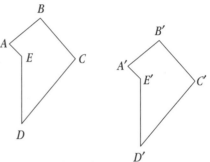

translational symmetry   A design has translational symmetry if can be created by copying and sliding a basic shape in a regular pattern. Translational symmetry is found in wallpaper designs and tessellations. The design below has translational symmetry.

Index

Index